MINDSET MECHANICS

The Science of Business Success Based on the Laws of Physics

Carolette Wright

Mindset Mechanics

The Science of Business Success Based on the Laws of Physics

All scripture quotations, unless otherwise noted, are from the King James Version of the Bible.

Visioneering Strategies, LLC

Lake Charles, LA 70607

For permission requests, speaking inquiries, and bulk order purchase options, email support@carolettewright.com

ISBN: 978-1-955562-04-1 (Paperback)

ISBN: 978-1-955562-03-4 (Hardback)

Cover Design by Abir Hasan

Edited by Carolette Wright

Disclaimer: The author, in no way, shape, or form, considers any of the information in this book to be advice, promise, guarantee, warranty, or any form of professional advice; it is intended for informational and educational purposes only. The results produced by the author or anyone referenced in the book are mentioned for illustration purposes only and are not intended to imply or suggest that you will have results that are at all similar to those in the book.

Printed in the United States of America.

Praise for Carolette Wright

"Mindset Mechanics grabs you from the beginning. The intriguing and compelling argument that businesses behave with the laws of physics is enough reason to start reading. The fact that Ms. Wright shows us exactly how to use these laws to get ourselves out of the status quo and move our businesses forward is the reason to keep reading! This book is intelligent and thoughtful in it's exploration of how the universal laws of physics can be applied to business. While my years in consulting have proven to me that our systems are intertwined and that we must take that into consideration as we grow, this is a new and dynamic exploration of this principle. I wholeheartedly recommend that anyone starting or growing a business reads this book!"

- Carrie Rickert, CEO Nomis Advisors

"As a visionary leader and high performer myself, I can appreciate the work of other ambitious and exceptional people like Carolette. Her mission to help leaders, business owners, and organizations build better processes and performance tactics to improve their delivery and especially help them reduce the risk of burnout is much-needed support these days. I highly recommend learning about her work and strategies to help people."

- Lizet Zayas, CEO of Successfully Business and Executive Coach

"Your creative thinking is very sound; it ignites you to think beyond... Your past experience gave you a lot of insights as to what is needed within the confounds of our Society... Great prompting!!"

- Carolyn Willis, Professional Licensed Counselor

Carolette has an uncommon insight and keen understanding of the mindset that business owners and leaders need if they want sustainable, healthy businesses and/or organizations. This is a seminal work that challenges our conventional thinking about the inner workings of business development and even shifts the perspectives of how we define success in business. (Hint: it has to be about more than making money!) There are so many insights to glean from, but she's not just pitching theory: everything she presents is actionable. From the first several pages of the introduction, the reader has a foundation upon which to start making meaningful pivots that will lead them to sustainable growth and help them to leave a lasting business legacy.

- Sheryl Landry, Founder
The Infinity Initiative

"The Spirit of the Lord God is upon me; because the Lord hath anointed me to preach good tidings unto the meek; he hath sent me to bind up the brokenhearted, to proclaim liberty to the captives, and the opening of the prison to them that are bound."

Isaiah 61:1

CONTENTS

DEDICATION

I dedicate this book to all the organizational leaders, managers, volunteer leaders, entrepreneurs, and small business owners who feel stuck in the details of building their businesses. Progress may be slower than expected, with some feeling stuck for months, even years, whether trying to start, grow, or pivot toward new opportunities.

This book is for those who have achieved some success but now face a standstill. It's also for the risk-takers who dedicate their time, energy, and resources to creating jobs, opportunities, and support for others.

I hope this book brings light in times of darkness, offers hope in despair, and empowers you to see the strength within yourself. My commitment to you is unwavering: I'm here to help you unlock your potential, achieve global influence, build wealth, and create your dream legacy.

Let *Mindset Mechanics* be your compass to embrace your dreams, strive for the best, and see no obstacle as impossible or adversary as unbeatable. Remember, you already have the tools for success—your potential is limitless. And if there's one takeaway you gain from this dedication, remember love always shows the way!

ACKNOWLEDGMENT

This book offers organizational leaders a complete system to ignite inner drive, accelerate business growth, and achieve personal and professional fulfillment.

Integrating principles from organizational leadership, classical physics, and metaphysics has shaped my understanding of leadership, personal growth, management, decision-making, and lifecycle development. Influenced by Bishop Wayne Malcolm, Dr. Ichak K. Adizes, and Lex Sisney, my approach draws from diverse perspectives to enhance leadership and growth.

My dream of business ownership began nearly 20 years ago, sparked by Dani Johnson of First Steps to Success. Sunny Lenarduzzi showed me how to establish authority in the marketplace, while Danielle Leslie taught me how to connect with different customer types across the product lifecycle. Figures like Myron Golden and Rabbi Daniel Lapin, combined with project management principles, have demonstrated the power of a clear, structured business approach. This book also draws inspiration from notable figures in classical and quantum physics, biology, and physical science, including Einstein, Lorenz, Feynman, Darwin, Newton, Carnot, Planck, Bohr, and others.

I am passionate about helping others and witnessing their success, and I am grateful to contribute to thriving businesses worldwide. I owe a heartfelt thank you to my children, whose encouragement has been unwavering, and to my family for their prayers and wise counsel. To my supportive circle, thank you for sharing your goals and insights.

And to my husband, you have seen my potential, even when I couldn't. Thank you for being my champion. I am blessed to have such amazing people in my life, and I owe everything to you!

INTRODUCTION

This book is for driven, goal-oriented, and ambitious leaders—whether you're an entrepreneur, small business owner, executive, manager, supervisor, or volunteer leader—working to move your organization from startup to growth. It tackles the question: *How can I drive my organization's success?*

As an organizational leader, you face relentless time pressures and energy drains. You need strategies that make a real impact quickly, even when it feels like an uphill battle to build wealth and secure your legacy. Between navigating a volatile market, managing tight budgets, expanding market share, and balancing work-life demands, it's a lot to handle.

With so many responsibilities, it's crucial to identify the link between actions and outcomes—the seen and unseen factors shaping your organization's path. You're aiming for stability, quick profits, and purpose-driven results. Yet, you may find your current plans aren't delivering, your team's dedication is slipping, and urgent action is needed to save your business.

It's overwhelming. You feel stuck, exhausted, and frustrated, sometimes even tempted to give up and leave everything behind. But deep down, you know that's not what you truly want. You started this journey to build wealth, realize your vision, and make a difference.

By now, you've likely tried various business theories with little success. You're tired, your business is financially stretched, and you're in need of a real solution to bridge the gap between where you are and where you want to be. *Mindset Mechanics* offers that bridge—giving you a clear, physics-based approach to unlock your organization's full potential.

You Did Everything Advised

You've tried expert advice, embraced new perspectives, and applied your best thinking to guide your organization. But despite your efforts, you haven't found the success you seek. You're always searching—balancing future goals, current management principles, and the pulse of your business. Yet, the path forward often remains unclear.

How can you confidently steer your organization toward success? How do you know when to trust your experience and intuition—and when they might be holding you back? What's the best strategy for your unique situation? And how can you reignite your enthusiasm while optimizing performance?

To answer these questions, you must identify the root causes of your challenges, just like a mechanic diagnosing an engine. It's not enough to treat symptoms; you must view your business as an interconnected system. Once you understand this, you can address the real issues and prevent future setbacks. That's what this book is about.

Mindset Mechanics offers a comprehensive system for improving your organization's efficiency, using a scientific approach grounded in first principles from physics and real-world experiences. I'm confident this approach works, not because I've followed traditional business theories but because it's collaborative, top-down, and based on systems thinking. I've applied universal principles that make sense in any industry. This book will help you diagnose problems, optimize performance, and achieve lasting success.

The key to success in business is following first principles.

Each person has their own beliefs about effective management strategies for organizations. Not all strategies are successful, and fewer are suitable for specific business ventures. This book clarifies misunderstandings caused by previous strategies and provides solutions for establishing a thriving business with raving customers and passionate employees. Embracing first principles thinking is the key to business success.

Applying physics to your business enhances performance, leading to greater success and growth. It instills confidence in implementing impactful strategies and unlocks opportunities. Physics is an indispensable tool for business success.

What is Physics?

In physics, first principles are the foundational elements that govern matter, energy, motion, and their interactions across the universe. These principles shape our physical world—and they also apply to business systems. Everything observable and measurable follows these fundamental laws.

So why use physics as an approach to organizational management? Because physics is universal, has its own language, and operates with a systems-based perspective. It offers a clear, structured way to understand and improve business performance. *Mindset Mechanics* uses these principles to bring clarity and effectiveness to management.

Physics is Universal

Physics offers a universal perspective essential for understanding how the world works. Alongside math, it serves as the language of science, with predictable patterns and laws governing everything. Physics shows us that energy is present in all things—and it can either propel you forward or hold you back. Now, imagine applying these principles to gain deeper insights into your organization's operations. Before moving forward, consider how integrating physics into your business approach could offer a powerful advantage. *Mindset Mechanics* provides the tools to make this transformation possible.

Physics Has a Language

Just as language shapes our perception, so does physics. Scientists are natural problem-solvers, but communicating with them can be tough due to language barriers. Physics, however, offers a universal framework that bridges this gap.

Applying physics principles in business enhances decision-making and problem-solving while attracting diverse talent and improving management. *Mindset Mechanics* uses this universal language to drive clearer, more effective solutions within your organization.

Physics Allows Us to Manage Systems

As we delve deeper, we find that organizations are often seen as Complex Adaptive Systems (CAS), defined by the Project Management Institute as dynamic networks that self-organize, adapt, and learn. This behavior is driven by individuals and teams, requiring human guidance to function efficiently. CAS evolve by interacting with their environment, which can shift their internal requirements over time.

Dr. Edwards Deming, a physicist, believed that organizations can't be fully understood by examining individual parts alone (what are often called "functions"). He emphasized the need to focus on how parts interact and once said, "You have to focus on interdependencies." This aligns with his Universal Systems Model (USM), which mirrors how nature manages systems. Leading organizations have found that physics principles provide a powerful framework for organizational success.

The purpose of *Mindset Mechanics* is to show how physics can elevate organizational performance, applying its principles beyond physical sciences to systems like businesses, families, and nations. As the digital revolution transforms industries, incorporating physics-based management, scientific methods, and technology becomes essential for staying competitive.

In this book, you'll discover physics-based models with a logical, natural approach to achieving business success. If physics explores matter, energy, motion, and their interactions across the universe, and management involves the principles, methods, and approaches used to guide organizations, *Mindset Mechanics* is the bridge of shared understanding that connects the two.

Physics is a management system that considers technology integral to the business landscape. Scientists and scientists at heart are the best problem solvers, but business owners are the best innovators. If we can get scientific professionals and organizational leaders to speak the same language, we can make astronomical advances throughout society. This book offers physics models that answer the question of *how to get your organization to perform successfully*—in ways that are natural, logical, powerful, and universally acceptable. By aligning scientific thinking with business strategy, organizations can solve complex problems, innovate more effectively, and thrive in a fast-evolving landscape.

9 Universal Principles of Mindset Mechanics

The *Mindset Mechanics* framework includes 9 Principles rooted in the fundamental laws of physics, designed to build successful leaders, cultures, teams, strategies, and processes. By applying these principles, you'll create lasting momentum, adapt quickly to change, and guide your teams toward success. If you're looking for solutions to drive growth and achieve your goals, physics holds the answer. Each principle is clearly explained, highlighting its benefits for organizational leaders and how it can transform performance.

Business Principle #1: Organizations are constantly evolving and changing.

Organizations are complex adaptive systems (CAS), evolving into distinct cultures and teams. Like individuals and machines, they self-organize, learn, and grow in response to environmental changes. As organizations expand or age, they continue to adapt and evolve. It's essential to understand that every part of the organization interacts independently and interdependently, driving ongoing evolution within the system and its surroundings. *Mindset Mechanics* helps you harness this adaptability to enhance growth and performance.

Business Principle #2: Organizations have attractive forces.

The *Universal Law of Gravitation* explains how larger masses attract smaller ones, with stronger attraction pulling things closer. Large organizations generally have stronger attraction in business, while smaller ones tend to have weaker pull. But this dynamic is changing—some small businesses have strong attraction forces, too. Organizations, like magnets, have fields of attraction shaped by their polarity. In business relationships, successful mergers and partnerships depend on aligned vision, values, culture, strategies, and decision-making. This polarity attracts the right leaders, talent, partners, and clients while repelling others. Most companies, regardless of size, often mismanage this magnetic force, unintentionally attracting or repelling against their best interests. *Mindset Mechanics* shows you how to align this polarity to attract the right elements for growth and success.

Business Principle #3: Organizations exhibit a clear state of balance and stability.

The *Zeroth Law of Thermodynamics* states that if two objects are balanced with a third, they are also balanced. In business, leadership and culture must align with teams and systems to enhance performance. Calibrating your organization to a clear standard of success is crucial. When leadership and culture are properly aligned, you can measure shifts and ensure balanced performance. Successful businesses maintain reference points and procedures that keep them in thermodynamic equilibrium—where leadership, systems, and internal dynamics are all in sync, resulting in optimal operations. *Mindset Mechanics* shows you how to achieve this balance for sustained success.

Business Principle #4: Organizations have standards and order.

The *First Law of Thermodynamics*, or the "BIG Energy Law," states that the total energy in a system must remain constant—it can't be created or destroyed, only transformed. In business, you cannot create energy; you must continuously seek new energy sources from your environment, ensure its quality aligns with your organization's needs, and process it into something useful and valuable. This law also suggests that energy becomes mass (stored energy), reflecting how vital resources can be harnessed for later use. Without sourcing fresh, high-quality energy, organizations risk breakdowns and failure. You cannot control your outcomes, but you can control your inputs. The better your inputs, the better your outcomes. For optimal performance, your organization's energy sources must be strong, consistent, and aligned with your goals. *Mindset Mechanics* guides you in managing these energy dynamics for sustained success.

Business Principle #5: Organizations inherently create chaos and confusion without proper management.

The *Second Law of Thermodynamics* states that chaos naturally increases over time, meaning all systems break down without intervention. In business, efficiency declines unless leaders actively invest in maintaining and improving their organizations. Leaders must preserve internal energy and subsystems—like processes, teams, and strategies to thrive. How you manage these energies will determine whether your organization thrives or just survives. *Mindset Mechanics* teaches you how to channel energy effectively to keep your business strong and adaptable.

Business Principle #6: Organizations with superior standards draw in exceptional talent and resources.

Materials Science, which blends various sciences, including physics, offers insights into the properties and structure of organizations. It shows how material choices—like energy, talent, and resources—shape culture and performance. Understanding materials science helps leaders make smarter business decisions for long-term sustainability. It emphasizes the need for criteria to assess energy, talent, and resources, ensuring they meet internal standards and adapt to environmental and market changes. Mindset Mechanics guides you in applying these principles to build a resilient, high-performing organization.

Business Principle #7: Organizations have a frequency.

Organizations build *frequency* through emotional intelligence and strategic communication. Emotional intelligence involves timing, tone, and authenticity. According to Harvard Business School, boosting emotional intelligence can improve team performance. When frequencies are misaligned, results suffer. Communication is the medium for frequency, fostering cooperation, participation, and commitment. Emotionally intelligent norms encourage positive attitudes, while alignment among leaders, teams, and clients creates a resonant and thriving organizational culture. *Mindset Mechanics* shows how to harmonize these frequencies for greater success.

Business Principle #8: Organizations can travel at the speed of light.

Organizations operate within Einstein's *Theory of Special Relativity,* where energy, like mass, can move at light speed. Similarly, organizations have mass and can drive swift change, but transformation requires time and investment. To outsiders, your organization's growth will seem sudden and remarkable. To achieve "light speed" transformation, everything must align—people, resources, and strategy—setting the stage for a bold leap forward. As the saying goes, "A thousand-mile journey starts with one step." You'll be amazed by the distance covered in the process. Physics provides a universal language that enhances problem-solving and decision-making. Integrating physics into management clarifies organizational dynamics and explains talent management and cultural dynamics. Applying these principles brings clarity to your approach and accelerates your organization's progress. *Mindset Mechanics* guides you in using physics to create meaningful, lasting change.

Business Principle #9: Organizations move based on laws of motion.

Newton's Laws of Motion offer powerful insights into managing organizational growth, change, and momentum. *Newton's First Law of Motion* explains inertia, highlighting how organizations either maintain the status quo or initiate movement. Inertia also affects how easily an organization can change direction. *Newton's Second Law of Motion* shows how net force, organizational mass, and acceleration drive change, emphasizing the forces needed for transformation. *Newton's Third Law of Motion* reveals the reciprocal nature of action and reaction, stressing the need to understand and leverage these dynamics for effective change management. Together, Newton's laws demonstrate the complex interaction between unyielding entities and unstoppable forces, providing a clear framework for navigating growth and transformation. *Mindset Mechanics* guides you in applying these principles to propel your organization forward.

Brief Summary

Even if you're new to physics, I've made the science simple, linking it to business concepts so you can quickly grasp the ideas and spot physics in everyday life. Understanding and integrating these principles will benefit your personal and professional growth, helping you attract the right people and resources.

Mindset Mechanics goes beyond traditional management theories by emphasizing real-world examples and physics principles that create real change, starting with you, the leader. By embracing these teachings, you'll boost your leadership skills, gain confidence, and develop a deeper understanding of leading effectively.

This book blends principles from classical and quantum physics with organizational leadership, offering a complete framework for growth. I've created clear definitions that show how these physics concepts connect to *Mindset Mechanics*, ensuring that you can easily apply them to enhance your personal and business success.

Here's a physics joke. While exiting the theater after seeing the movie *Star Wars*, a physicist bumped into a fellow physicist. Inspired by the film, he yelled to his friend, "May the mass times acceleration be with you."

The actual quote in the movie is, "*May the force be with you.*" In Newtonian physics, force equals mass times acceleration. So, please don't forget that many frameworks are available to teach organizational theory.

Each is equally impressive, but only *Mindset Mechanics* explains how physics works with forces that cause motion. Motion is about the internal forces—personal drive, skills, experience, capabilities, resources, and support systems that make any theory magical. If any theory suggests otherwise, it's wrong. The force is within you, and it has always been within you.

Finally, I would like to talk about the target audience of this book. The laws of *Mindset Mechanics* apply to a broad spectrum of circumstances and enterprises; however, this book is designed specifically for the organizational leader, small business owner, volunteer leader, executive, entrepreneur, manager, and whoever wants to use technology, including digital media platforms, artificial intelligence, and the internet, over the next 12 months to either (1) start or grow their own business and establish themselves as industry leaders, or (2) build a brand within a large organization and become a "vital person of influence." This targeting is for a variety of reasons:

1. With a scientific background, I coach entrepreneurial leaders, understanding their challenges and opportunities firsthand. My mission is clear: to help organizational leaders remove stress and burnout caused by ineffective performance strategies. My passion for guiding them fuels my desire to share *Mindset Mechanics* globally.
2. Startups, growth-focused companies, and personal brands need swift action and holistic strategies. Applied physics provides a methodical approach to help organizations learn, adapt, and execute quickly. Companies that master fast learning and decision-making rise to industry leadership while consistent small changes drive growth and results.
3. In today's fast-paced, tech-driven world, analyzing and synthesizing data quickly is critical for competitive advantage. My mission is to help startups, growth-based companies, and brands integrate technology effectively into their products, processes, and decisions.
4. Technology is most powerful when paired with a clear message. I aim to help leaders amplify their voices, fine-tune their businesses, and seamlessly adapt to every product iteration or market shift.

Mindset Mechanics is your guide to thriving in a complex, evolving landscape.

Whether you're part of a startup, a growth company, or simply want to improve your life or organization, *Mindset Mechanics* offers principles that make you a more effective person, manager, or leader. You'll see tangible positive changes by applying its laws to your workplace.

Pursuing fundamental principles that drive performance has always been a part of human progress—from early tools to the scientific method. As our understanding grows, we keep asking: *How do things really work?* This quest leads to continuous improvements in how we operate.

As you read *Mindset Mechanics,* I encourage you to focus on the organization you wish to enhance and ask: *How do I get my organization to perform successfully?* The answers are within these pages.

---

1. Landry, Lauren. "Emotional Intelligence in Leadership: Why It's Important." *HBS Online*, 3 April 2019, https://online.hbs.edu/blog/post/emotional-intelligence-in-leadership. Accessed 26 September 2024.
2. Sisney, Lex. *Designed to Scale: How to Structure Your Business for Exponential Growth*. Organizational Physics Incorporated, 2022.
3. Sisney, Lex. *Organizational Physics - The Science of Growing a Business*. Lulu.com, 2013.
4. LePore, Domenico. *Moving the Chains: An Operational Solution for Embracing Complexity in the Digital Age*. Business Expert Press, 2019.
5. Montgomery, Angela. "Physics and Management: What You Must Understand to Lead and Manage Today." *Intelligent Management*, 19 April 2021, https://intelligentmanagement.ws/physics-and-management-what-you-must-understand/. Accessed 28 September 2024.
6. Hansen, John W. "Universal Systems Model." *University of Houston*, 2001, https://www.uh.edu/~jhansen/6302SYL%20Fall%2001.htm. Accessed 4 October 2024.

HOW TO READ THIS BOOK

In the film *Atlas,* Jennifer Lopez's character, Atlas Shepherd, reluctantly connects her mind with a human-like artificial intelligence (A.I.) machine. Doubts about forming this bond stem from her past, but she ultimately embraces technology to protect humanity from annihilation.

Even though many people fear artificial intelligence's potential, it's clear that AI is becoming increasingly normalized and embedded into our society. *Mindset Mechanics* is your connection between the science and business worlds. It provides excellent value for those interested in increasing business intelligence, understanding business behaviors, improving organizational performance, and achieving desired results. It's best that you don't need a scientist to use it.

Mindset Mechanics is divided into two main sections. The first section is the "mindset" section, which gives you the behind-the-scenes background story that shares a little about who I am and how *Mindset Mechanics* came about. I share the twists and turns of my (slightly unusual) journey to where Mindset Mechanics came together. I hope my story provides you with three things. First, the context for what *Mindset Mechanics* is; second, a recognition of opportunities where *Mindset Mechanics* can be used; and third, perhaps the most important, inspiration when things are tough. Every successful entrepreneur's story is filled with near failure. My story is no different, and if you're like most people, I think you'll be surprised to hear how accepting the possibility that almost losing my life gave birth to *Mindset Mechanics.*

The second section, the "mechanics" section, consisting of four parts, shows you what the Mindset Mechanics Interface (see **Figure 1**) is and how to apply it in your business through a comprehensive framework. For example, if you seek assistance in recalibrating your business vision, the answers within Section I contain the insights you need. Should you encounter difficulties concerning talent management, you will find Section II to be a treasure trove of wisdom on handling strong nuclear personality types.

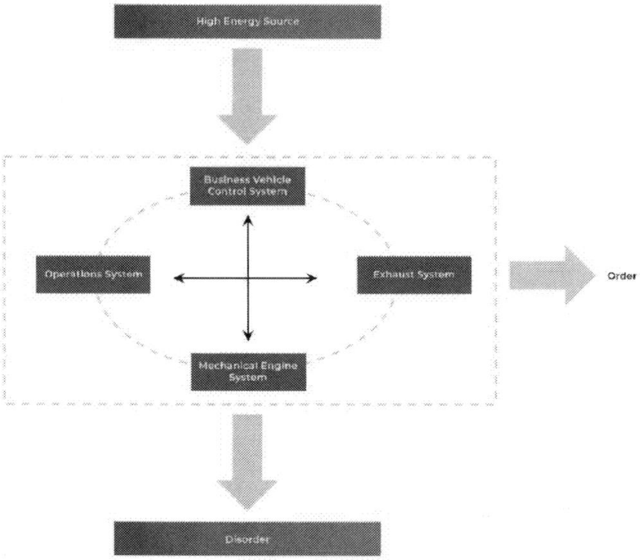

Figure 1. Mindset Mechanics Interface

If you like to skip around, you can go straight to the "mechanics" section, and it's all there for you. Of course, you can always return to Section I to find out how I came up with *Mindset Mechanics*; doing so may make more sense and add context. The choice is yours where you begin. No matter where you start, this book is designed to be comprehensive and interactive. My goal is to help you understand the concepts and strategies of *Mindset Mechanics* and give you the most up-to-date business application of these principles at the same time. For this reason, you can unlock the bonus section by visiting the *Mindset Mechanics* website at **https://themindsetmechanic.com** to get exclusive access to resources designed to complement this book.

Your Mind is a Powerful Machine

Machines are becoming more like human computers each year, and the evolution of the automobile is a prime example of *Mindset Mechanics*. Modern vehicles are essentially computers that make decisions, yet they still need mechanics for repairs and humans to operate them.

To understand the *Mindset Mechanics Interface*, consider two metaphors:

1. **First, imagine your mind as a powerful computer**—capable of learning, adapting, and replicating itself. Think of it as having a 3D printer connected to a vast neural network with the potential to create your dream team, design strategies, and attract your most valuable paying clients.
2. **Second, picture your business as a machine, with you as the operator.** Like any automotive machine, it requires the right design, components, and systems to reach your destination. The goal is to manage the journey smoothly, maximize performance, and deliver your products and services safely and efficiently.

The *Mindset Mechanics Interface* serves as your business's operating manual, connecting the computer (your mind), the machine (your business), and you, the operator. With this framework, you can run your business as effectively as a mechanic maintains a vehicle—driving toward success with precision and ease.

Your Mind is the Control Center

If you are passionate about reaching your business goals, realizing the leader's mind is the control center for your organization is essential. We will create structure and delve into the five critical aspects of understanding and utilizing the **Mindset Mechanics Interface** (**see Figure 1**).

1. Like everything else in the universe, your organization operates within a system governed by the laws of physics. Success depends on how well-aligned all components are.

2. The mindset you develop directly impacts your organization's culture, leadership, teams, and processes.

3. You are in full control of your machine's operations, driving its success.

4. Your machine requires fresh energy to maintain momentum and produce results. The quality of your planning, resource use, maintenance standards, and operating practices will determine

your system's success.

5. *Mindset Mechanics Interface* as a system helps you harness these principles to optimize performance and drive growth.

Mindset Mechanical Interface

The *Mindset Mechanics Interface* maps out your organizational system represented within the dotted lines. From the top, it receives new energy from higher sources; at the bottom, it releases lost energy; and to the right, it produces results. Like our previous metaphors, your organization is a supercomputer with visualization, networks, performance, and essential applications. Similarly, an automotive machine has an operator, fuel, an engine, and exhaust systems.

Your business functions like a vehicle, with interconnected subsystems interacting with the environment. The dotted box within the *Mindset Mechanics Interface* represents the whole organization—its components, leadership, and operations team. This box extracts energy, creates order, and releases waste, much like a vehicle does. It includes an operator, systems, parts, and the flow of information and resources, all working together to drive success.

Business Vehicle Control System

The *Business Vehicle Control System (BVCS)* consists of the operator's mindset and the machine's functionality. It provides mental frameworks and strategies successful leaders use to differentiate themselves and think clearly and purposefully. It asks, "Who are you?" and "Who do you want to be?" *BVCS* drives the entire organization by using sensors to detect conditions, qualify the quality of information, and send information; a control unit of leaders to process information; and actuators to make strategic decisions and carry out commands. The *BVCS* is where the supercomputer for the leader and organization sits, but like all computers, it must be connected to a power source. Your power source provides constant energy.

The *Business Vehicle Control System (BVCS)* serves as your organization's control center. It houses the Laws of Thermodynamics: conservation and disorder, the cultural hub, and reflects the overall system's collective mindset, directly reflecting its leadership. Focusing on the organization's inner workings, aligning subsystems, and positioning them in the right direction is essential to achieve significant progress. Alignment requires focusing the organization's culture, structure, processes, and people on common goals to drive the business forward. Once the internal systems are aligned and force is applied in the correct direction, changing the organization's course becomes easier due to the momentum built up. Therefore, organizational expert Lex Sisney describes this leadup to organizational growth as establishing the business vehicle's "mass" and its direction from the outset of any "motion."

Operations System

The *Operations System* is the hardware and software systems used to control subsystems and operate various features and functions of the organization, such as providing steering power, navigation, safety, and maintenance. The *Operations System* focuses on managing functions of the organization, including hardware resources and activities; controls operations like steering, switching gears, braking, and fuel consumption; and uses software to support safety-critical systems.

The Operations System hardware leverages the four fundamental forces of the universe to steer the organization forward: *Strong Nuclear, Weak Nuclear, Electromagnetic,* and *Gravitational* Forces. These forces come into play depending on your team, the nature of the work, and the tasks needed to create products or complete processes. People forces drive all actions and their effects, helping organizational leaders shape and adapt to environmental changes. The Operational System software converts the forces into personality styles: *Creating Innovation, Guarding Implementation, Nurturing Unification,* and *Connecting Communication.* These personalities help you understand how to manage people and safely interact with your employees and customer types.

Mechanical Engine System

The *Mechanical Engine System* (MES) converts fuel energy into mechanical energy, making organizational movement possible. The MES focuses on the strategies to supply power to your organization's engine. Your business vehicle needs to be near a viable fuel station or have the capabilities to convert available resources into usable energy. The energy needed to generate power is sourced from the environment. New energy is required for your business systems to continue functioning and producing work. Aligning the right fuel energy type and quality impacts your business vehicle's performance. *Mechanical Fuel System* (MFS) depends on implementing the right actions for fueling. The *Mechanical Systems Lifecycle* (MSL) depends on making the right decisions at the right time to maintain synchronized systems.

The Mechanical Fuel Systems (MFS) enable businesses to produce work independently of time, relying solely on displacement. A company's success depends on aligning the *business model, business products, business market,* and *business execution* for seamless product and service delivery. Mechanical Systems Lifecycle (MSL) predicts the energy quality needs of the business vehicle as it matures and ages.

Exhaust System

The *Exhaust System* measures the condition of the output of an organization as energy flows through the system. The Exhaust System determines the overall health of an organization. It helps determine if critical business vehicle subsystems are aligned or misaligned. If the business vehicle is misaligned, mechanical servicing is needed to check the cause of the undesired output. First, a *Mindset Check* is performed, which consists of a visual, eight-point preventative maintenance inspection to locate possible leaks in the system. Then, a *Mechanics Check* is performed, which connects the entire business vehicle to Mindset Mechanics, which acts as a diagnostics machine to help you collect information, monitor, and regulate whether organizational subsystems are at risk of failure, and determine their performance.

The Mindset Mechanics Interface serves two purposes: to minimize disorder and maximize performance. Disorder, or undesired outputs, represents the fuel energy from your business vehicle's surroundings that cannot be used for productive work. Performance represents the desired outputs created by an aligned and healthy organizational system.

You Control Your Business Vehicle's Performance

What if everything in your business reflects your actions and decisions? If you're not satisfied with the quality of your business performance, adjust the inputs, align the components, and you'll change the outputs. If you're not satisfied with the direction your business is taking, find a new path. Many organizational leaders fail to realize that how your business functions is a direct reflection of the control center—your mindset. All the accomplishments, ideas, and results of your business are created by you. The Mindset Mechanics Interface gives you complete control. It guides you through the management and maintenance processes. If there's something you don't like, you have the power to change it.

Key Takeaways

This book aims to empower you to become a mechanical operator who can evaluate yourself and your organization, assist your team, adapt to your environment and circumstances to achieve your goals and maintain high-performing systems. What's the business vehicle's internal temperature? Are the main components and subsystems supplied with the necessary resources, such as proper maintenance, fuel energy, fluids, or internal energy? Are you receiving the renewal and support you require as a leader? Have you asked yourself and your leadership team what drives them? Are you an innovator who can recognize patterns, anticipate the next wave of change, and envision new possibilities to attract resources? Are you a nurturer who values daily involvement to help others improve their skills, adapt to new training, and embrace personal development? Are you a protector who enjoys getting involved in the action to accomplish tasks and thrives on seeing tangible results? Are you a coordinator who enjoys organizing and bringing people together to create execution plans?

Second, after reading this book, you should be able to assess the hardware that forces your organization to propel forward and the software that provides the management style your team needs to win! Does your business vehicle have the necessary mass to maintain stability, attract organizational attention, overcome friction forces such as drag and inertia, and build momentum to push past obstacles and make impactful decisions? Are all its components and subsystems agile enough to handle rapid changes and address emerging issues? Are your people, processes, mission, vision, and values aligned and prepared for accelerated motion? Are your internal forces tailored to meet the organization's needs and priorities? Is the right talent in the correct positions, and do you have the proper people mix to facilitate organizational development and growth? Have you considered whether you are applying the appropriate force to your organization's mass and whether you have the proper business structure to align with your culture, vision, and values?

Third, after reading this book, I want you to build your capabilities to print anything you desire and create your organization’s path. The only way to achieve your *wildest* dreams is to have a continuous supply of new energy with the best quality and highest efficiency. Even with the best *business model* to fuel your business vehicle, your organizational system must have the capabilities to shift gears through *business execution* and *lifecycle* changes and manage its resources after energy is extracted from the surrounding environment in the form of money, resources, partnerships, and relationships, and create *business products* and services that a *business market* filled with your most valuable prospects (MVPs) will benefit from and enjoy.

Last, after reading this book, I want to inspire you to align your business systems because this helps you materialize anything you can envision and carve a way forward. Maintaining a consistent source of fresh, high-quality energy with maximum efficiency is the key to making your most ambitious aspirations come to life. Your approach should be capable of overseeing the processes of acquiring resources from the environment, such as finances, materials, alliances, and connections, of developing your offerings, penetrate markets to reach your most valuable prospect or most valuable payer (MVP) and carry out sales.

Think again about your business vehicle. As a mechanical operator driving down the *Road to Success*, it's crucial to ensure your car is adequately prepared for the conditions and environment of the *Road to Success*. This means selecting a business vehicle that is aerodynamically equipped to minimize external drag forces and ensuring that the *Mechanical Engine Systems* can handle the *force* of weight from the *mass* of the organization and the team. Mass and force are key components to reaching higher accelerations.

A skilled mechanical operator must (1) monitor fuel consumption, (2) prioritize engine speed compared to the vehicle's condition and load, and (3) also observe vehicle gauges for signals determining how the internal systems are functioning. Neglecting these signals can lead to unwanted forces that deteriorate the system, overconsumption of energy, decreased performance, and untimely vehicle failure. A failure to monitor, control, and maintain your vehicle could leave you and your passengers stranded, missing in action, or stuck on the side of the Road to Success.

Business ownership and entrepreneurship are not for the faint of heart. You have to deal with constant challenges that require you to be on top of your toes; fluctuations in the market can make things extremely volatile and unpredictable; unexpected obstacles pop up, and your business may start to lose its competitive edge, which may be hard to retain.

The principles of *Mindset Mechanics* are about your mindset as a leader and your mechanical skills as a mechanic. *Mindset Mechanics* is interested in implementing strategies and cultivating a mindset to identify root causes, leverage internal and external forces, and change underlying factors to alter your performance results. You can control, manage, and maintain your business vehicle to produce any desired result. It's not difficult to apply *Mindset Mechanics* to your organization. If you follow the steps in this book in the prescribed order and implement them at high standards within the next 12 months, it's possible for you to create the organization of your dreams.

According to social and cultural researcher Sara Redshaw, who has spent years studying the effects of driving and vehicle mobility among inexperienced drivers, she believes that vehicles represent freedom, independence, and individualism. Vehicles allow you to travel wherever and whenever you want, with whomever you choose. The same is true for your business vehicle. This book aims to equip you with the mindset and knowledge to select, construct, and steer your business vehicle as a responsible mechanical driver. You'll know how to navigate down the Road to Success, reach your desired destination, and safely steer yourself and others to freedom. Are you ready to journey in a well-designed, highly efficient, and fueled business vehicle as a competent mechanical operator? Let's get started. Enjoy.

---

7. Ferriss, Timothy. *The 4-Hour Workweek.* Harmony/Rodale, 2009.
8. GlobalData. *Cybersecurity: who are the leaders in automotive operating systems for the automotive industry?* Our report provides a holistic view of the automotive industry's Automotive operating systems landscape, covering a wide range of topics such as innovation landscape, market readiness and company deep dive for select innovators. no. GDDT-AR-2989-MP-L1, GlobalData, 19 October 2023. https://www.just-auto.com/data-insights/innovators-cybersecurity-automotive-operating-systems-automotive/?cf-view.
9. Halderman, James D., and D. Mitchell. "Steering & Suspension - Automotive Technology." *LibGuides*, 30 March 2023, https://wiregrass.libguides.com/c.php?g=1035978&p=7530246. Accessed 26 September 2024.
10. Landry, Lauren. "Emotional Intelligence in Leadership: Why It's Important." *HBS Online*, 3 April 2019, https://online.hbs.edu/blog/post/emotional-intelligence-in-leadership. Accessed 26 September 2024.
11. Sisney, Lex. *Designed to Scale: How to Structure Your Business for Exponential Growth.* Organizational Physics Incorporated, 2022.
12. Sisney, Lex. *Organizational Physics - The Science of Growing a Business.* Lulu.com, 2013.
13. "The Standard Model." *CERN*, https://home.cern/science/physics/standard-model. Accessed 26 September 2024.

PART I: Developing a Growth Mindset

CHAPTER 1: This Crisis Saved My Life

I had just finished our 2022 business tax filings, and everything seemed to be going well—both personally and professionally. But suddenly, the bright future I envisioned vanished.

On Tuesday, January 10, 2023, after seven sleepless nights with only fifteen-to-twenty-minute nap and a loss appetite, I found myself in a predicament. I was running on sheer adrenaline. By 7 PM the next day, an ambulance rushed me to the hospital. I felt like I was having an out-of-body experience, watching everything unfold as I was strapped down. I asked to be taken to a military hospital, but my request was ignored, leaving me suspicious of those around me.

Something's Not Right

I was suffering from sleep deprivation. I engaged in conversation, eyes shut, nodding my head, saying, "I know this looks weird. But my eyes feel extremely tired. I am not sleepy and can't fall asleep." The irony wasn't lost on me; I had spent the last twelve hours completely unaware of what awaited—a brief break in mental stability.

As I sat on my living room floor, I lost all sense of time, caught in a meditative sway. Flipping through a photo album, my fingers traced the protective film over the images as if I were reading Braille. While my eyes were closed, I felt disconnected. Without warning, I finally drifted off into sleep—but only for a moment—as I was still swaying and unaware of how much time was passing.

When the paramedics arrived, they asked me what was wrong. I simply said, "*I'm just tired. I need to sleep. All I want is to sleep.*" But despite my exhaustion, deep sleep continued to elude me. My sister called for an ambulance, alarmed by my unusual state. It was unsettling for her—and even more so for me. I'd never experienced anything like it.

Let's pause for a moment. I realize I've skipped over many details, and you're probably curious about what led to this crisis. I need to provide additional context for you to understand the turmoil taking place.

Science of the Mind

Years ago, I purchased a book about brain function out of curiosity. I had no idea how valuable this book would be after my crisis. In *Stroke of Insight* by neuroanatomist Dr. Jill Bolte Taylor, she shares the complexities of the mind vividly explored through her personal stroke experience. In it, she explains that effective communication requires a shared reality where nervous systems mirror each other's perceptions. People must sense and interpret information similarly for genuine connection, with a common reality shaped by a common language and shared actions.

Our brains are in constant flux, connecting and disconnecting new neurons over time. The neurons you have today have been with you since birth and remain unless damaged. What's remarkable is that these connections are shaped by your unique experiences. Two people can experience the same event yet walk away with completely different interpretations and beliefs. These differences arise based on how we process past experiences, make sense of new ones, and filter information—positive or negative. Processing happens rapidly and subconsciously without our awareness of what's taking place.

Communication necessitates a shared reality.

Dr. Taylor emphasizes the warning signs of how the mind processes data during a mental crisis. I discovered that sleep deprivation has several harmful effects on the mind. Here are three major symptoms signaling the onset of severe sleep deprivation psychosis:

1. Perceptual distortion
2. Delusion
3. Hallucination

I had lost touch with reality. Confused and struggling to recall details, I clung to false beliefs, convinced others were plotting against me. Shadows seemed real, and my perceptions felt distorted. The crisis altered my experience, making my reasoning feel both rational and confusing, with an overwhelming sense of mental hyperactivity.

Hemispheric Asymmetries

I lost track of time and created elaborate stories. In the right hemisphere of your brain, time doesn't exist—only the present moment. The right-brain collects data independent of time, viewing moments as timeless and abundant. It thrives on intuitive, out-of-the-box thinking, embracing spontaneity, creativity, and limitless possibilities. It's the source of our artistic flow—free from inhibition or judgment and thinks in pictures.

The right-brain gathers data independent of time. The left-brain measures data and divides time.

I couldn't tell the difference between true and false. The left hemisphere of the brain operates very differently from the right. While the right-brain embraces spontaneity, the left organizes details methodically and strings moments together, creating a sense of time—past, present, and future. The left-brain measures, categorizes, and processes data, holding your language center to describe and communicate everything logically.

Evening Break

The night before the ambulance ride, I didn't see my fears as delusions or hallucinations—I saw it as a spiritual battle. I prayed throughout the night, wrestling with fear. My imagination spiraled into dread, picturing demonic spirits in my children's rooms and fearing the worst for my husband, stationed miles away in New York City. I had never experienced such intense fear and prayed for strength.

As dawn broke, a sense of calm came over me. I had made it through the night, and the fear had lifted through prayer. With renewed resolve, I got dressed, dropped off my kids, and headed to the BJACH Hospital in Fort Johnson, Louisiana.

Guardian angels protected me.

As I drove to Fort Johnson, fear crept in—I felt as if other drivers were out to hit me. The mere thought of reckless drivers made me wary. So, I drove slower for safety. As I drove, I could not help but feel as though guardian angels were watching over me.

Reserve negative judgment over other people's experiences.

I arrived at my appointment without incident, but when the nurse handed me a survey, I hesitated. At that moment, I remember how a former colleague had once shared how completing this survey led to her involuntary admission to a behavioral health facility, leaving her feeling trapped and confused. I hadn't fully understood her experience then, but now I felt the weight of it. It's true—you can't know someone's struggle until you're in their shoes.

I voiced my concerns about the survey to my physician before transitioning to the purpose of my visit.

Despite societal progress, mental health remains stigmatized. I'd seen extreme cases in movies like *Girl, Interrupted*, *A Beautiful Mind*, or *Rain Man*, but nothing prepared me for the broader reality. My colleague never fit the stereotype, so I hadn't connected mental health to her situation.

Seeking medical care can feel vulnerable, especially when it involves mental health. Comparisons and contrasts complicate things further, making self-advocacy essential. But feeling safe and educated is equally crucial.

During the appointment, I believe my doctor sensed something was off, but I concealed my severe sleep deprivation by working hard to maintain my control. The rest of the visit felt surreal and sluggish. I struggled to process information and jotted things down just to keep up.

When I returned to my car, fear crept back. I knew something was seriously wrong but couldn't grasp it fully. I kept telling myself: *What are you doing? Drive home safely, and rest when you get there. Go slow, pay attention, and you'll be fine.* I repeated this mantra, trying to stay focused as I drove home.

Answers in Stillness

The ambulance brought me to my local hospital. Surrounded by quietness, I drifted in and out of sleep while a sense of calm washed over me. What felt like days passing was only hours. But during this time, vivid dreams or visions emerged, revealing that the relentless pace I had been pushing was doing more harm than I'd admitted. As I lay there, unable and unwilling to fully wake, a new fear crept in—was this leading me toward a different reality? Would a mental health facility become my final destination?

In my state of insomnia, my inner spirit was unsettled. I saw a future that felt all too real: I was confined to a mental hospital, trapped in my own mind, unable to communicate. I even envisioned my husband's new wife—she was familiar but happier, healthier, and freer—the version of myself I longed to be.

These visions felt like my personal hell, not one of "brimstone and fire," but of isolation and loss of control. Lying there, unable to move, I learned a hard lesson: I had been neglecting what truly mattered—my family—for the sake of ambition. Exhausted and hospitalized, I realized the importance of valuing life and appreciating the present.

Strong is overrated.

Being strong is overrated. Choose courage instead. I surrendered to my circumstances, accepting whatever fate had in store, and found gratitude in my family's well-being. In a prayer, I prayed, "If this is my fate, I won't complain. I'm grateful. I take full responsibility, accept the consequences of my actions, and find peace knowing my family will be safe and happy—even if I must remain here forever.

Make a Decision

In the midst of the vision, I couldn't recall any spiritual songs or biblical verses to reassure myself. But when I surrendered, I heard the words, *"Just have faith."* Instantly, fear left my body. A confident voice followed: *"You get whatever you want. It's your choice. Just decide what you want, and it's yours."* I chose life. I didn't want this to be my fate. I realized I needed to change—my old ways of thinking and pushing myself to the limit had to end.

The next morning, I woke up relieved, thinking it was just a dream or a warning. Feeling better, I believed I was going home to make amends. But to my surprise, the hospital had arranged my transfer to a behavioral health facility in Baton Rouge. I wasn't informed until the ambulance arrived. I had requested Fort Johnson, but they sent me to Baton Rouge without explanation.

The vision had prepared me for this moment. Though I was frustrated, I wasn't afraid. I stayed calm, accepting the situation even without clear answers until days later.

The Baton Rouge Facility

At the facility, they gave me a new gown, socks, and toiletries before escorting me to my room. I asked for shoes and was told they'd be brought to me.

I quickly realized I was in for an experience, and my A-type personality kicked in. I decided to document each worker's contact with me—they needed to know I wasn't an average patient. I noticed two types of workers: some seemed confused about my presence, while others were visibly struggling with their own issues. The level of professionalism and treatment of patients was poor.

For two days, I asked for shoes but was ignored. By the third day, my feet hurt with every step. It was time to lay down the law, so I spoke firmly to the staff, telling them that once I left, several would be terminated. Minutes later, they found a pair of oversized men's shoes. It was a small victory—those sandals felt like foam pillows.

I realized the staff couldn't see past their biases, treating patients as defenseless and incompetent. Thankfully, my amazing family brought me proper clothes and shoes and helped to shift their perception. As staff started seeing patients in regular clothes, I noticed they started treating them differently, and their behaviors changed.

The Medication

On the first night, I asked the nurses for a natural sleep aid—melatonin—the only medication I felt comfortable taking. The nurse said she was waiting for the pharmacy to release it, along with any prescribed medication.

I am not the type of person who likes taking lots of medication—even for a migraine, I'll wait until it's unbearable before taking Tylenol. This medication was no different; I weighed the risks before making a decision to administer it. Regarding this facility, I didn't feel safe or trust their decisions. For one, they failed to properly explain what was happening or provide me with the side effects of the medication. So, I insisted on administering what I was most familiar with. I wanted to see if melatonin would allow me to sleep while maintaining alertness. For my entire stay, I declined their offer for stronger medications. I didn't want to get too comfortable in an unfamiliar environment and wanted to be home to rest.

Over the next three days, my attempts to sleep were rudely disrupted by loud noises, staff arguments, and horseplay, leaving me restless and irritated. By the third day, I was over it, and I was told I could go home. One counselor asked if I wanted an Uber or family pickup. To save on fuel and inconvenience to others, I chose the Uber. Wrong choice. That Uber never arrived. I doubted it was even ordered.

Once it was clear what I was dealing with, my family got even more involved, but I was inconvenienced because it was a holiday weekend, which meant I had to say three additional days, and two more days were added as they worked out the details for my release. As I struggled to rest in the facility, my team still needed me to run the business, which remained a priority, even though I needed to rest. It was this experience that reinforced my need to change how I lead. I was a functional CEO. I needed to become more vital and equip my people to make decisions to keep things going in my absence.

The Fire Alarm

On the sixth day, as we returned from lunch, the fire alarm unexpectedly blared throughout the building. The staff quickly ushered patients into a holding room while they unlocked the courtyard to lead us outside. The situation seemed exaggerated, with anticipation filling the air.

While we waited for them to open the courtyard, I stood by a window, praying to go home and for my spirit not to break. Suddenly, one of the counselors stopped in front of me. With a swift gesture, he signaled for me to follow him. It felt cinematic. I weaved through the other patients, turned a corner, and reached him. Amid the blaring alarm, he said urgently, *"Get your things, quickly. I'm getting you out of here."* I was ready—I stayed ready. I had kept my suitcase packed, sensing the moment of my departure was always near.

I gathered my things calmly but quickly. Once outside, safely in the car, my family explained the chaos of getting me discharged—largely due to my refusal to take the medication. My family helped to release me when earlier the doctors I was told they'd keep me there indefinitely unless I conformed to taking the prescription.

In those six days, I witnessed troubling realities: unprofessional conduct, patient mistreatment, mixed communication, mental manipulations, and was presented with unrecognizable meals they called food. Yet, I felt no malice toward the staff. It was clear they, too, were struggling—temporary patients in a system that seemed to trap everyone, much like guards working in the prison system. The only difference was they were able to go home while the patients were subjected to their system. It was also clear that I was supposed to experience this crisis. I needed to be a witness to help change the stigma and be a voice to the voiceless in hopes of improving things.

The behavioral system is broken and overwhelmed. I believe many people working in the healthcare system are trying to help, but in many cases, some do more harm than good. Witnessing the sheer number of patients and their constant and urgent needs leaves little room for these workers to express compassion or patience. I couldn't ignore my experience—it revealed the deep crisis and need for fixing system leaks.

My situation was avoidable, and I've since committed to teaching organizations how to remove unnecessary stress in doing business and learning the mechanics of business so they can achieve their goals while finding fulfillment. My crisis was a blessing in disguise, as it helped me to see that now's the time to address the issues associated with leadership stress and business burnout head-on.

---

14. Taylor, Jill Bolte. *My Stroke of Insight*. Viking, 2008.

CHAPTER 2: Working Hard to Achieve Dreams

When I was around 11, Monopoly became a family favorite. My uncles and aunts playfully teased each other as we learned about wealth and decision-making. The game taught us how to buy and sell properties, seize opportunities, and collect rent. Success required strategy, and those who mastered the rules had a real chance of winning.

Around that time, one of my uncles, still in high school, moved in to live with us. To make extra money, we started a candy business together. We sold popular brands like Reese's, Snickers, M&Ms, and Skittles—$1 for regular packs and $2 for king-size—at school during the day and door-to-door at night. It was a profitable venture, supporting my uncle and earning us extra cash. But what truly motivated me was the process: acquiring valuable items, marking them up, and finding eager buyers.

Humble Beginnings

Our lifestyle was simple. We never lacked essentials, but some things were out of reach. Our car wasn't the latest model—my mother drove a black two-door sedan. My three siblings and I shared one bedroom with two sets of bunk beds—a cozy and efficient setup. We didn't have the latest fashions and often shopped at Walmart or the local Goodwill for basics like shoes, jeans, and church clothes. But we always had what we needed: full stomachs, a safe place to sleep, and a home filled with laughter and contentment. Still, I noticed what the world offered, and I wanted more. I was determined to be successful.

Looking back, school came easily to me, perhaps because I believed education was my only path to validation and attention. My mother always encouraged us, saying, "You have a choice to go to school and get an education. Do better than I did." I took that to heart.

As the oldest of four siblings—two younger sisters and a brother, plus another brother from my father's side—I often felt a sense of responsibility to set an example and strive for more.

More Than Skin Deep

Colorism affected me as a young child. Whenever we met new people, they always commented on how "lovely" and "light-skinned" my younger sisters were. I understood that I was "dark-skinned" and rarely received compliments about my appearance. As a little girl, comparisons mattered, and I learned that dark skin wasn't always seen as "pretty." But I did have one quality that stood out—I was smart.

Realizing this, I decided to channel all my energy into my studies. I consistently excelled, ranking at the top of my class from elementary through high school.

Going to school as a kindergartener is a blur. But I vividly remember every year afterward. In the second grade, I was already writing in cursive. By fourth grade, I often finished assignments faster than my classmates, who would look at me in awe and ask, *"How do you write so fast?"* Reflecting on this brings a smile to my face. They had no idea that, for me, it wasn't just about speed. I was driven by a fierce sense of competition. In my 10-year-old mind, I was competing against the entire class. Along with getting excellent grades, finishing first with a perfect score was the ultimate win. I developed mental shortcuts and strategies to absorb principles, process information, and implement quickly.

In second grade, my family moved to Minnesota to visit family. I was excited to meet my maternal grandmother. However, the stay was brief, and we returned to Nashville before I started the third grade.

The Fourth Grade

My mother's friend enrolled me at Dodson Elementary School on the first day of school. Without hesitation, he confirmed my age and informed the school that I was a fourth grader. I was the same age as his younger sister. While enrolling in the fourth grade, I knew I was scheduled to attend the third grade, but I didn't correct them. For the last 4 years, I'd wished for the day to be skipped ahead a few grades—but today, I'd settle for being placed in my correct grade. So, they assigned me to my fourth-grade class, and I was finally pleased to have found a solution for my *"T-1 Disaster"* (more on that later).

As the days progressed in the fourth grade, having only completed the second grade, the work was not difficult for me, and I excelled. A few weeks into the new school year, my exuberance over my performance got the best of me, and I was quickly humbled. When the teacher casually remarked, "We are going to review what you learned in the third grade," I couldn't contain my excitement. I raised my hand and declared, "I didn't go to the third grade."

As soon as the words escaped my lips, I looked at my teacher's face, and a wave of dread washed over me like a tidal wave. My heart sank as I realized that my admission's consequences meant it was over. That day, I was shifted to a third-grade class, and the devastation was palpable. Why couldn't I have just remained silent? I was reminded of my tendency to "talk too much," which finally backfired against my plan.

One Year Behind

I have always been competitive. It was my way of proving to the world that I was one of the best. So, when my peers often asked, "Why are you a year behind?" It was hard for them to understand when I explained. I never knew the full answer until many years later.

I was born in Saint Louis, Missouri. This is where I spent the first four years of my life. After a visit, my mother relocated our family to Nashville, Tennessee. She finally found an overnight job; none were available in Saint Louis. As my kindergarten year approached, she asked family members to help me onto the school bus each morning. I don't know why I don't remember much about kindergarten; my memory wasn't fully developed, or I rarely attended. But I do remember a deep affection for school.

Even as a toddler, I was advanced and encouraged to learn. My mother potty-trained me before I could even walk. I was told I held engaging conversations with adults as a toddler. In preschool, I was taught to count in Spanish, and it wasn't long afterward that I read books before elementary school.

In 1987, I was going into the first grade, and I was excited. But when we arrived, there was a delay. My mother was instructed to meet with the guidance counselor; as I listened, I noticed my kindergarten year was being discussed, and in a moment, my excitement for first grade disappeared.

T-1 Disaster

I recall the moment when she examined my record from the previous year. She concluded that I wasn't prepared for first grade. Not for my grades—I had straight A's—but because of my poor attendance record.

It was confused. No one asked me for my opinion. I was prepared and could read at a second-grade level. She never inquired about my abilities; I was never assessed. Instead, the guidance counselor decided that since we were a "poor" family, I would be placed in the Transitional First (T-1) program. I was fully aware of their discussion and felt a surge of disappointment. She was persuasive enough to convince my mother that I was not being held back and T-1 was the most suitable choice for me.

Transitional First (T-1) emerged as a Title I initiative crafted by the Department of Education specifically for children transitioning from kindergarten to first grade. This program became a component of the Elementary and Secondary Education Act of 1965, a crucial strategy in President Lyndon B. Johnson's campaign against poverty. Rather than repeating kindergarten, they decided to enroll me in this program. When we finally reached my new classroom, it was toward the middle of the day, and the students were getting ready for their afternoon naps.

I have glimpses of memories from my kindergarten days. One vivid detail is that I never took naps during class. While my classmates—whom I affectionately dubbed "babies"—drifted into their dreams, I would lie down, watch the clock, and count the minutes until the nap break ended. It was a curious paradox that the only time I ever succumbed to sleep in school was in T-1, a place that marked the beginning and end of something significant in my young life. The day came when I was finally roused from slumber, the last to awaken. I can still hear my teacher's gentle voice breaking through the fog. "Carolette, wake up. You must have been tired." If only she could fathom the storm of emotions swirling in my six-year-old mind—a mixture of disappointment and confusion stemming from the choices laid out by the guidance counselor.

As I sat there, a single thought consumed my mind: "Why am I stuck in T-1, and how can I change this?" Deep down, I understood that I was grappling with a situation that, being a kid, didn't help. The odds that the adults would listen were not in my favor. Determined to defy expectations, I set out on a quest to prove everyone wrong, clinging to the hope that, just perhaps, I could be skipped up to the first grade, where I belonged. The harsh reality of the education system, driven by funding, meant I remained in T-1 for the entire year. After watching the reactions from the first graders who knew me in kindergarten, I was careful to keep my age to myself, aware of the teasing that suffocated those who had been held back. I refused to let their narrative of failure define me—my story was yet to be written.

In the face of numerous disappointments, I persevered and thrived each year, all thanks to the strong foundation my mother established for me. She dedicated herself to my early childhood education, carefully guiding and nurturing me in a way that my siblings, who received less of her time due to her parental responsibilities, could not experience. Balancing her efforts to ensure we were all taken care of, she worked tirelessly. Being the firstborn allowed me to spend time with her, and her investment in me shaped who I am today. Through her dedication, she ignited within me a deep desire to learn.

Become an Engineer

I excelled in art, math, and science, but math truly captured my heart. It resonated with me in a way that felt almost instinctual. Unlike English composition, there was only one correct answer, and it was a straightforward path to uncover through formulas. I was in the band during my freshman year of high school, and we played at the graduation ceremony. While giving their speeches, I discovered that the valedictorian and salutatorian would be pursuing engineering. Their selection interested me deeply. Eager to learn more, I searched the internet, American Online (AOL), and Vista Search, and entered keywords to understand what it meant to be an engineer. I thought if the top students were aiming for engineering careers, I should find out why.

I discovered that engineering blended art, mathematics, and science. From that moment on, I had a clear sense of direction for my career. Every class I took and every book I read focused on shaping my future as a mechanical engineer.

Selecting a College

The first time I heard about *Historically Black Colleges and Universities* (HBCU) was when I received a recruiting brochure in the fall of 1999. It was from Howard University, the only HBCU institution I considered and visited. I asked the adults around me and only heard great things, so I decided to visit. During my time on the campus, it became clear to me that this was where I was meant to be. Before this visit, the opportunity to attend was almost missed.

First, I received a rejection letter from my top choice, Vanderbilt University. Once I knew I wanted to attend Howard University, I turned down an offer to attend Rose-Hulman Institute of Technology in Terre Haute, Indiana. Then, I received my decision letter from Howard University's College of Engineering, Architecture, and Computer Science (CEACS). The opening line went something like this: "We are sorry to inform you that you were not accepted into the incoming class of 2004." At that moment, my hopes shattered into a million pieces; I didn't have a Plan B. I was disappointed by another setback—but something could not accept that "no."

In an unexpected moment of clarity, I called the Administration Office and asked why my admission was declined. I discovered they had old test scores. I gave the woman on the phone the updated scores, and she accepted me into Howard University but enrolled me in the College of Arts and Sciences. She advised me to complete my first semester and transfer to the College of Engineering, Architecture, and Computer Science. I don't know how different my life could have been if I hadn't made that long-distance call to Washington, DC.

Arrival to Campus

College was the first time I truly felt challenged in school. I dedicated myself to my coursework, putting in extra hours to ensure I kept up with the demands—I worked hard. I also spent hours in the financial aid office, seeking assistance to help cover the cost of tuition and room and board. I didn't know if each day was going to be my last. I grappled with the possibility of receiving a letter instructing me to vacate the dormitory due to my unpaid balance. Strangely enough, that letter was never delivered. Meanwhile, I watched as other students in similar situations were evicted from the dormitory, withdrawn from classes, and returned back to their hometowns. Miraculously, I was spared.

Entrepreneurship & Leadership

At Howard, the engineering school frequently invited recruiters and consultants to provide career guidance and discuss job opportunities. During my junior year, one group of consultants arrived to share entrepreneurship. This was the first time I considered owning my own business as a career path. I knew this was my path, but I didn't know how to start.

My college hired adjunct professors to help teach major courses, including subjects such as Energy, Heat Transfer, Systems and Controls, and a few other subjects. They played a pivotal role in igniting my passion to specialize in Energy.

Going to school in Washington, DC, had its advantages. I participated in the Consortium Program, a cross-campus partnership that allowed students to take courses for credit at other universities. To graduate on time, I took advantage of this program by enrolling in classes at George Washington University and the University of Maryland. This opportunity enriched my education and academic journey in unimaginable ways.

I've always thought that one day, I may run for office. I have always been on the student council, and undergrad was no different. I was elected vice president of the College of Engineering during this period. In addition to this honor, I was accepted into the esteemed Howard University Science, Engineering, and Math (HUSEM) Undergraduate Research Program. My time was further enriched as I was voted Chapter Secretary of the National Society of Black Engineers (NSBE) and took on the role of Treasurer for the American Society of Mechanical Engineers (ASME). Moreover, I completed my duties as a resident assistant, each role extending its own unique opportunity for me to lead, share my passions, articulate my visions, and manage my educational finances. Leadership was an inherent part of who I was, a magnetic force drawing me toward responsibility. My ability to lead became my lifeline whenever there was a need for direction. I stepped forward, frequently chosen by my peers to take the reins.

As my graduation from Howard approached, I found myself in a predicament familiar to many seniors: I was eager to pursue my passion for leadership while also making the most of my engineering degree. But I had no job offers lined up, was undecided about continuing my master's education, and didn't know how to start a business.

The Last Mile

In the summer of 2004, I enrolled in a significant course offered at the University of Maryland through the consortium program. I could not find financial aid before the deadline, so the university dropped me from class. Determined to graduate in the spring of 2005, I spoke to the professor and asked to complete the class despite my situation. He agreed.

I visited with the assistant registrar the next semester and explained everything to her. I didn't know what to expect, but I had faith. What happened next was unprecedented. On a rare opportunity, the university agreed to receive late payment and retroactively credit me for completing the class. A wave of relief washed over me when the University of Maryland agreed to reinstate my enrollment and allow the professor to record my final grade.

The spring of 2005 brought challenges in my last semester. I had taken another course through the consortium and informed the professor that I was graduating and needed to make arrangements to complete my final exam early. Thankfully, he was accommodating. He permitted the entire class to complete their final exams ahead of schedule. However, my final grade was submitted before May 1st, causing disappointment. After five difficult years of studying, I was about to miss my goal. This cut was deep, reminiscent of Stevie Wonder's *Rocket Love.* The lyrics capture the experience of a person taking on an exhilarating journey with someone they love, only to be abruptly dropped back down to the realities of life.

You took me riding in your rocket; you gave me a star,
But at half a mile from heaven,
You dropped me back down to this cold, cold world.

Indeed, it felt as though I had been lifted up, cast aside, and left in a state of uncertainty. With only a couple of weeks before graduation, I informed my family to cancel their trip because it seemed unlikely that I would walk across the stage.

While the graduating seniors buzzed with anticipation and joy, I found myself on the outskirts, unable to join the festivities. The absence of this crucial grade meant I wouldn't even receive my cap and gown. I would only be a spectator in a celebration that should have been mine.

Feeling as if I had reached the end of my rope, I approached one of my professors after class and shared my predicament. Honestly, I didn't expect anything. I had already surrendered to the situation. I only intended to inform my supporters why I wouldn't attend graduation. However, she had a different reaction. She sprang into action. She contacted the University of Maryland professor and reached out to the administration at Howard University. Before I knew it, I received news that my final grade was submitted. I would graduate and walk across the stage. What a relief!

Once again, my prayers were answered. In an act of surrender, I found myself being showered with blessings. My intense desire to graduate had built a barrier between my mind and what I thought I needed. I harbored thoughts of scarcity—what I may lose, what I lacked, what I had given up. But once I let go of my will, what I wanted came to me effortlessly. It would take me years to truly grasp how mindset and science principles conspired on my behalf and worked in my favor.

CHAPTER 3: The Unexpected Plot Twist

After completing my engineering degree, I embarked on a new chapter of my life, moving from the bustling streets of Washington, DC, to the quieter atmosphere of Pittsburgh, PA. Just three months into this new adventure, I celebrated another milestone – my marriage. As my husband and I began to lay the foundation for our family, the urgency to enter the workforce began to weigh on me. I quickly recognized that honing my interviewing skills was essential. Throughout the following six months, I landed numerous interviews, each opportunity serving as a valuable learning experience, but I was often overlooked in favor of another candidate.

In the meantime, I still needed to earn a living, so I took a job at UPS during the holiday season. I quickly realized that the fast-paced environment and demanding physical labor were not for me. Seeking a slower pace, I accepted a sales position at Sears Outlet. Truthfully, I found little joy in that role. There was one particular salesperson whose personality clashed with mine. His competitive nature was annoying, and he didn't play fair. Instead of vocalizing my frustrations, I learned to stay focused and sidestep his antics, hoping that one day he would calm down and learn to collaborate with me and our other co-workers. But I didn't plan on staying long enough to witness his change. Fate intervened, and I received two job offers.

The first offer was for a supervisory role as a team advisor. In this role, I would lead a team of 25 individuals within the food manufacturing industry. I was also offered an entry-level reliability engineering position with another company. I believed the team advisor role offered me the best opportunity for growth, whereas the reliability engineer position didn't provide enough room to advance as a leader, and it was less lucrative.

In December 2005, I accepted the team advisor position with Schrieber Foods, Inc. They gave me a generous sign-on stipend and took care of all relocation expenses from Pittsburgh, PA, to Clinton, MO. They also assisted me with four weeks of transitional housing. I felt I made the right decision.

In December 2008, I left Schreiber and entered a new role at General Mills. By 2011, I took an eight-month break to pursue entrepreneurship before returning to the workforce and joining Kraft Foods in July 2012.

With each new position, my salary increased. I took on greater responsibilities, faced increased complexity, and invested more of my time into my career. That's when it dawned on me. It was my pursuit of higher pay that added complications to my life. I realized of my ambitions, interests, and skills, I would never get paid or have the role I wanted unless I created it. Thus, I abandoned my climb up the corporate ladder and redirected my energy toward launching my own business.

After I graduated from college, my heart was filled with the ambition to launch my own enterprise. Fast forward seven years, and after welcoming my second child, I found myself standing far from those early aspirations. The dreams I once held seemed to have drifted further away, leaving me discontented and unsatisfied, as my job lacked the purpose I yearned for.

In 2014, I left Kraft Foods, seeking a fresh start by returning to school to refine my goals. I was convinced that further education would pave the way to my aspirations. Meanwhile, my husband made the significant decision to enlist in the Army. His commitment impacted our family dynamic and gave me the precious opportunity to dedicate time to my children, pursue my Master's degree, and continue my business journey.

In 2015, I accepted a position with a federal contractor. Two years later, I transitioned to another federal contracting job that was more aligned with engineering. Eventually, I moved on to accept an engineering role with the Louisiana Department of Environmental Quality (LDEQ). At night, I worked on myself and my business.

This was the rhythm of my existence for a decade and a half following my college years. My days were filled with work, and when I returned home, I dedicated my evenings to acquiring new skills, scouring the internet for promising opportunities, or providing services to clients. Countless times, I encountered the urge to throw in the towel, grappling with the sensation that my life was slipping through my fingers. Yet, each moment of hesitation was met with an unseen force that propelled me onward. I had developed an insatiable craving for hard work, and the thought of being idle filled me with a sense of failure. Through it all, I discovered a silver lining—every effort was purposeful, and time was never squandered.

Every lesson I had absorbed, every endeavor I had undertaken, and everything I had achieved in the realm of business was intentional. woven into a grander scheme that would begin to unfold for me in the summer of 2021. I delved into website development, explored the nuances of online marketing, practiced cold calling, managed invoicing, honed my customer service skills, and navigated the complexities of employee and contractor relationships. My knowledge felt overwhelming; my experiences were like scattered puzzle pieces. I found myself trapped in an employee mindset, unsure of how to unify all this knowledge into a cohesive business that ignited my passion. I possessed the necessary skills and talents, yet what I truly lacked was the fervor that would bind my focus and dedication to a meaningful mission.

Life has a way of unfolding differently than we anticipate, and my journey is no exception to this truth.

When I interviewed for my position at Schreiber Foods, my new boss was on vacation. Her absence meant she hadn't planned on bringing anyone new aboard without having any input on the decision. During our first encounter, she clarified her feelings, almost saying, "I didn't guarantee you anything. So, what do you expect me to do with you?" I didn't feel welcomed.

I found myself in a position where I had to prove my abilities. The initial months were quite challenging, as she seemed determined to make me give up. No matter what I did, she had a way of questioning my every move. In her frustration, she took her complaints to the plant manager who had brought me on board. He was fond of me, and I genuinely appreciated working under his guidance.

As time passed, she began to open up to me, and we gradually forged a friendship. She took on the mentor role, understanding my aspirations for greater responsibilities. Given the state of the economy, she chose to equip me with the skills necessary for my upcoming promotion. Shortly thereafter, Bob accepted a different position within the company, and she chose to return to her hometown in Minnesota.

In 2007, as I awaited the arrival of my first child, my ambitions for a corporate career began to wane. I decided to embrace entrepreneurship, setting aside my previous goals. Though I diligently performed my work duties, keeping my BlackBerry close and maintaining a professional demeanor, I felt a growing disconnect from the idea of climbing the corporate ladder. I often questioned whether this was all there was to working life. A sense of unfulfillment settled in, and I worried that my precious moments with my newborn would be compromised. However, the necessity of earning a living compelled me to stay in a job that brought me little joy while searching for a different career path that might bring more satisfaction.

In 2008, as I navigated the job market, I stumbled upon an opportunity at Schreiber Foods in Antioch, TN, that caught my eye. However, the company had paused relocation assignments while waiting for the economy to stabilize. This reality pushed me to consider my situation; I needed to find something closer to Nashville, TN, my hometown. After the interview, I accepted a more lucrative offer with General Mills in Murfreesboro, TN. If I had to be employed, I might as well be near my roots and surrounded by family.

It's important to grasp the situation at that moment. I had reached a point where I was pretty disillusioned with the work-life culture.

However, when Schreiber Foods discovered that I was collaborating with a different food manufacturer, there was a chance they might request the return of my BlackBerry. Yet, to my relief, that concern turned out to be unfounded.

While at General Mills, I launched a property preservation company that proved to be highly profitable, yet it left me feeling exhausted. I juggled the demands of working the third shift—from 11 PM to 7 AM—with managing work orders for employees and contractors. In between, I cherished moments with my son, often sacrificing sleep. This relentless routine persisted for over a year, and I operated on empty. Fortunately, our business's flourishing success allowed my husband to transition to full-time work with the company while he cared for our son. Although this shift offered me some much-needed relief for building a profitable business, it didn't alleviate my guilt for not being as present in my son's early life as I desired.

Once our house in Clinton, MO, sold, we moved into a beautiful luxury apartment, and everything seemed to fall into place. My job and the business we were running brought in nearly six figures. Yet, the industry's culture left me feeling disheartened. It was an environment to which I didn't see myself changing, nor did I want to devote my life. I was frustrated as contractors repeatedly overlooked instructions, forcing us into the daunting task of reworking properties. Furthermore, I watched as employees neglected to document their work adequately, leading to chargebacks and payment delays that only added to my stress.

On an ordinary day, I turned to my husband and expressed my struggle: I couldn't keep up with the business any longer. Little did I know then that I was already five months along with our second son—a realization that struck me intensely, especially right after we had just completed a major painting contract. In hindsight, I found myself wondering whether all the hard work had really been worth it. I was feeling completely overwhelmed, and it struck me how I had invested so much of my energy into my job while being oblivious to the fact that I was pregnant.

During my unwavering quest for financial stability, I unexpectedly found myself amid a crisis. Internally, I was crumbling. Each day, I managed to push through, yet the heaviness of my depression hovered over me like an ominous cloud that I couldn't quite see. Even before reaching the age of thirty, I began to sense the clear indicators of burnout sneaking in, a realization that struck me as both shocking and sobering.

After Barack Obama's 2008 election, I found myself tuned into the Trinity Broadcasting Network, captivated by a speaker named Dani Johnson, who graced the screen for several consecutive days. Usually, the lineup consisted of guests deeply rooted in religious themes. However, Dani brought an entirely different energy to the stage. A Christian millionaire, she boldly recounted her inspiring journey from poverty to prosperity. Her charisma was magnetic. She radiated warmth; she interpreted the principles in a manner that resonated with everyday experiences. As I listened, it felt as though she was the manifestation of how I saw my inner self.

I've secured my place at the upcoming First Steps to Success Conference in Houston, TX, and before starting my new position with General Mills, I ensured they understood my need for the week off to participate in this vital event. As I reflect on March 2009, I wish I could confidently say this conference transformed my life for the better. However, the reality is quite different.

Dani completely transformed my perspective. I encountered a woman who radiated strength and confidence. A thought crossed my mind: "Who does she believe herself to be?" My ingrained beliefs were struggling to stay hidden. Before me, a woman was pursuing a path I recognized as my calling. A wave of divine jealousy surged through me.

I traveled to the conference to expand my knowledge and gain valuable insights for my business. However, what unfolded was a journey of self-discovery, as I found myself reflecting on the person I could evolve into.

Dani addressed every facet of human existence, offering a pathway to success that became achievable if we embraced clear guidance and adhered to fundamental principles. It was a pivotal moment, as my understanding of psychology faced its first true test. When I walked away, I felt cornered and confronted by the weight of my realizations. Yet, there was also a sense of clarity—I recognized the extensive effort required to dismantle my ingrained misconceptions. A vast expanse of knowledge lay before me, highlighting the gaps my engineering degree had left untouched. It became evident that I had to embark on a fresh journey of growth and self-discovery.

Once the conference wrapped up, I found myself wandering into Barnes and Noble, ready for my customary book hunt. I loved to meander through the aisles, eager to discover hidden gems among the shelves. This day, I spent around half an hour browsing when a title suddenly caught my eye: The Four-Hour Workweek by Tim Ferris; its allure is impossible to resist.

In 2009, I stumbled upon a book that was unfamiliar to me, yet it had already gained a reputation in the business world. At that time, I hadn't even heard its name. Although I didn't find every idea within its pages agreeable, the book sparked a transformative shift in my mindset. To my amazement, I started using it as a roadmap for my journey. I meticulously analyzed its contents and wrote notes in the margins, eager to unveil the deeper insights lurking within. My thirst for knowledge drove me to seek out YouTube channels, acquire additional books, and grab hold of every online resource that I believed could aid in my quest for entrepreneurial growth.

Later that year, in 2009, I embarked on the journey of starting a property preservation business. This venture kept us busy until 2011. Rather than shutting down the operation, I chose to adapt. I gathered all business documents, invoices, photographs, and contacts collected over the previous two years and transformed them into an eBook.

I created online tutorials, hosted engaging webinars, and offered personalized coaching sessions. With determination, I taught myself the intricacies of building a website, and when I finally brought it to life, everything began to shift in my favor. The prospect of "making money on the internet" felt surreal, almost like a dream come true. It was as if I had stumbled upon a hidden treasure, and the excitement of it all was exhilarating. This moment marked the discovery of passion and purpose.

As I immersed myself in this work, a sense of empowerment began to wash over me. My enthusiasm ignited a desire to expand my efforts, yet a frustrating barrier stood in my way: I lacked the necessary skills to swiftly turn my ideas into reality. Tasks that I anticipated would only consume a few days of my time stretched into weeks, leaving me disheartened and sapping my motivation as I waded through the endless details. Instead of steering my business towards growth, I found myself buried in the daily grind, and the experience was exhausting. Yet, amidst the struggle, a fire of determination burned within me, refusing to let me quit. To fuel this newfound resolve, I decided to attend a Robert G. Allen conference in Anaheim, California, all in the hopes of enhancing my online presence and propelling my business forward.

After returning home, brimming with innovative ideas and fresh strategies, I was thrilled to see a significant change—it felt like my sales had tripled overnight. A few months prior, I took a leap of faith and decided to leave my role at General Mills, eager to pour my energy into my growing business. However, as soon as I stepped away from the safety of my job, I was hit with an unexpected wave of stress that stifled my creative flow. Although the online revenue I was generating was sufficient to cover our living costs, it was far from the stability I craved. With that, apprehension crept in. I began to question the very methods I had once wholeheartedly embraced. Instead of spending 8 hours working on my business, I was not working the business 16 hours a day. This was worse than having a job. The thought of returning to the workforce lingered in my mind. The responsibility to help provide for my family weighed heavily on my shoulders. I had my engineering degree to lean on, and the thought of a steady paycheck grew more enticing by the day. Ultimately, in July 2012, just eight months after I had left, I started the interview process and received eight job offers at the same time. They were competing for me. I had grown as a leader, and taking time off made me a better communicator. I decided to follow the money and accepted a position with Kraft Foods.

Only two weeks into my new position, a profound realization settled heavily upon my shoulders—I had made a mistake in my decision. I should have stuck with working on my business. It was as if a fragment of my very essence was gradually fading. Nonetheless, fueled by a steadfast dedication to my family, I decided to persevere through the challenges ahead for the next two years, embracing this sacrifice for their benefit.

Beneath the surface, turmoil swirled in my mind, unseen by those around me. To the outside world, I was flourishing—well-educated, successful, and surrounded by a loving family. My life appeared flawless, and I took immense pride in the quality of work I engaged in. However, an unsettling void tugged at my heart. Whenever I attempted to share my recent revelations and experiences with my loved ones, their puzzled expressions hinted that they viewed my dreams as nothing but a desperate chase for wealth.

The reality was quite different from what others perceived. An intense desire for prosperity simmered deep within me—a desire not born. from self-serving goals but rather from a sincere wish to help others in my circle of influence. However, I struggled to find a way to direct this passion. No matter how hard I tried, I could never quite articulate the core of my entrepreneurial spirit to anyone. My friends and colleagues never seemed to understand why I consistently walked away from high-paying jobs. Maybe they viewed my actions as somewhat irrational, but I had a bigger plan. I wanted a thriving career, time freedom, and happiness.

2014 I embarked on a new chapter by deciding to part ways with Kraft Foods. This pivotal choice marked the beginning of a journey focused on self-discovery. During this time, my husband stepped into the role of the breadwinner, allowing me the freedom to explore my true passions and seek out a career that would bring me genuine joy.

CHAPTER 4: Standing on the Shoulders of Giants

Before I left Kraft Foods, I decided I needed to slow down in order to speed up. While enrolled in my master's program, I immersed myself in YouTube University. Every time someone recommended a book, I would immediately add it to my shopping cart and later purchase it off Amazon.

I found myself ordering books, courses, and various entrepreneurial resources, gradually transforming into an information enthusiast. I was on a relentless quest for answers, often at the expense of my finances. During the period from 2008 to 2014, the online world felt like a frontier full of chaos and opportunity. While many strategies turned out to be worthless, a few did capture my attention. Danielle Leslie taught me how to create and market my online course. Sunny Lenarduzzi showed me how to build my authority on YouTube and become the sun in my solar system. But from the start, there was only one speaker who truly resonated with me.

I found myself drawn into Jim Rohn's world, becoming a devoted student of his teachings. He introduced me to an array of remarkable books, effectively enriching my personal library. It all began between the ages of two to four, when my mother and teachers sparked a passion for reading within me. From that moment, I sensed a deep connection between knowledge and entrepreneurship. When I first heard Jim Rohn declare that "readers are leaders," it resonated with me on a profound level. His engaging style and sharp wit captivated me completely, compelling me to transcribe his every word into longhand notebooks—a task that was no straightforward endeavor. Working from home, I immersed myself in his videos day and night, dedicating every spare moment to absorbing the insights of influential speakers, diving deep into the intricacies of business, and transforming my mindset.

As clear as day, I can hear Jim Rohn's voice:

- *"Don't say, 'If I could, I would.' Say, 'If I can, I will.'"*

- *"Don't wish it was easier; wish you were better. Don't wish for less problems; wish for more skills. Don't wish for less challenge; wish for more wisdom."*
- *"How long should you try? Until."*
- *"Formal education will make you a living; self-education will make you a fortune."*
- *"Success is nothing more than a few simple disciplines practiced every day."*

Listening to Jim Rohn became part of my daily routine for the next three months. I encountered many inspiring speakers like Leslie Nicols, Les Brown, Robert Kiyosaki, Myles Monroe, and others during this time. I delved into copywriting, public speaking, effective communication, marketing, and coaching. Gradually, I came to understand the vital role of authenticity in speaking. It became clear that the audience cared more about my genuine interest in their issues than my intelligence or degrees.

One of the most important lessons I learned was how to improve my communication skills and find my voice. There wasn't any secret hack or hidden trick; it all came from finding my passion and doing some soul work. I was encouraged to pay attention to my emotions. What got me excited? What sparked that fire inside me? What truly helped me go from sad to glad? If something I said, read, or stumbled upon could pull me out of a tough spot, then I knew it would do the same for others. As a leader, I started taking notes on myself. I realized I was onto something, so I dedicated the next few years to being more aware of what mattered to me. I began to embrace my story, even the parts that brought shame. It was time to face my past, regain my power, and see them as life lessons.

Fast forward to 2021. During the COVID-19 pandemic, I decided to kick things off and start a trucking business. Everything I'd picked up over the past 15 years was put to the test. I grew the company to five straight trucks and my first million in the first year! I shifted my mindset from thinking like an employee to an entrepreneur. I figured out how to create a business that helped others. I ended up with a team of over 12 employees. We grew quickly. My background in management, leadership skills, and strong values kept everything on track.

My values acted like a GPS, guiding me whenever I hit a crossroads. They helped keep our business genuine and on the right path. In the end, it was my strong leadership and principles that helped me navigate through it all. After each hurdle, I always found myself back on top. Everything I had learned from all the books, webinars, courses, and videos was finally coming together. I was standing on the shoulders of giants.

CHAPTER 5: Taking the Leap of Faith

Sometimes great things are created from unexpected and unlikely places. That couldn't be truer for my first million-dollar business, which ended up laying the groundwork for *Mindset Mechanics*. The unlikely place? A trucking company.

As the pandemic continued to unfold, the idea of starting a trucking company lingered in the back of my mind. However, I found myself held back by the fear of fully committing to a single venture. Following my experience in the property preservation business, I was hesitant to invest my energy into something that might end up making me feel trapped, especially if it turned out to be a business I didn't truly love. Deep down, I recognized my potential, but my selectiveness kept me from taking the plunge. Inspired by the phrase "multi-passionate entrepreneur," coined by Marie Forleo, I identified with it entirely. I was blessed with a multitude of passions, yet I struggled to settle on one, particularly since none of these interests were currently generating any income.

I often reflect on how I didn't really pick trucking; rather, it picked me. It was during the summer of 2021 when one of my brothers expressed a desire to become a truck driver. He secured a position as a subcontractor for a prominent third-party logistics firm. The subcontractor manager offered him a chance to start his own business as a subcontractor. Filled with enthusiasm, he set out to find the funding needed to embark on his new journey. Seeing this opportunity's investment potential and nature, I decided to buy a truck and take on the contract under my business name. Since I was located in Louisiana and the opportunity was based in Nashville, my brother stepped up as the on-site representative for my company.

In just a few weeks, we had secured the necessary authorization for the truck to transport freight, adhering to the regulations set forth by the Federal Motor Carrier and Safety Administration (FMCSA) within the Department of Transportation. As we eagerly awaited the commencement of our contract, we thought it prudent to take the truck on a journey across the open road. Though I hesitated, we decided not to take the truck to the dealership for an oil change. Instead, we entrusted a local mechanic with the task. However, disaster struck just 100 miles outside Nashville—the truck began to stall. After two long days, it became apparent that the engine had run dry of oil, resulting in significant damage. I was heartbroken. After enlisting the help of a couple of tow trucks, we received a staggering quote from the dealership: over $35,000 for repairs, encompassing both parts and labor.

What followed took a leap of faith. I came across another truck that had low mileage. I presented the truck to my commercial loan officer and shared what had occurred, and with her help, we managed to arrange financing for the second truck. This experience taught me the importance of a proactive mindset when facing business challenges. By seeking assistance and being open to solutions, I was able to overcome obstacles and move forward with my plans. I was able to gather the resources needed to repair the first truck. Before I knew it, I unexpectedly became the owner of two trucks.

In the early days of my venture, the first couple of months saw me relying solely on my personal savings to keep everything afloat. However, after experiencing the costly setback of the engine failure, it surprised me how steep the expenses could be when running an asset-based business. This realization prompted me to think strategically about leveraging my credit. Before the unfortunate engine incident, I had already put the first truck up as collateral, paving the way for opening trade lines.

Once I built up my business credit, the arrival of the second truck proved to be a game-changer. It showcased my incoming revenue, making it a simple task to secure additional credit lines. With this newfound momentum, I calculatedly acquired as much funding as possible while the situation remained favorable. Similar to funding the cost of an engine replacement, I knew that if I ever needed financing, it might not be readily available. So, I actively pursued various lines of credit, such as business credit cards and other trade accounts.

It was months before I found utilizing any of those credit lines necessary. Looking back, I am thankful for the engine replacement. It gave me the foresight to secure funds in advance, ensuring that I was prepared for any challenges.

As opportunities to expand our fleet of trucks arose, our operations went from a single vehicle to two and soon grew to five. We secured numerous contracts across Nashville. My focus was deeply rooted in supporting my team, satisfying our clients, overseeing the maintenance of our trucks, ensuring compliance, and dealing with financial matters. Amidst all this, I became so immersed in the day-to-day demands that I lost sight of our overall revenue. It wasn't until I was getting ready to meet with our accountant that the realization struck me: we had grown into a million-dollar enterprise.

Before, I never could decide on what type of business to start. Shifting my focus to help someone else decide for me. Before I focused on myself, I could not take the leap, but once my choice was to help someone else, I could take the leap.

CHAPTER 6: When Things Come Together

I was selected to participate in the Fall 2023 Goldman Sachs' 10,000 Small Businesses Cohort 20. The journey hadn't been easy; I faced rejection in two previous interviews. I decided this would be my last attempt to pursue this prestigious opportunity. As the day of the interview dawned, I could feel a wave of confidence wash over me as I reflected on my business's achievements. I kept my emotions in check, projecting a serene exterior while remaining unfazed by the uncertainty of the future.

Participants learned about raising capital and creating a growth plan in Goldman Sachs' 10,000 Small Businesses (GS10KSB) Program. Like many entrepreneurs, I was resolute in expanding their businesses and ensuring success. Before the program kicked off, I found myself exhausted from the relentless grind, feeling a sense of battle fatigue. I have always believed that our life's work should bring joy instead of turmoil. In all its beauty, life is far too valuable to be weighed down by stress. Deep in my heart, I am certain that our true mission on this planet is to appreciate the gifts we possess fully and use these gifts for the good of others, yet too often, we become ensnared in outdated ways of thinking that keep us stressed and distracted. We're focused on the wrong things or doing the right things in the wrong order. This desperately needs to change.

The GS10KSB program facilitated a turning point when I finally confronted a painful truth: I was in the wrong industry. For too long, I was trying to make it work, and I had overlooked the signs and patterns in my life that were urging me to change course. By December 2023, I had made a firm decision—I would dedicate the rest of my life to following what truly inspired me. The monotony of dealing with transportation brokers had worn me down. I reached my limit, constantly juggling margins, funneling money to mechanics, and collaborating with teams that struggled to manage themselves. It was time for a new path that resonated with my passions.

My worst nightmares become my reality. I found myself ensnared, caught in a web of my own design. The business I had created stifled my ability to showcase my true talents. I had poured my heart and soul into this endeavor—fighting tirelessly, putting in endless hours, embracing risks, and sacrificing so much of who I was—that I couldn't help but lament the moment I first acquired my truck. It felt like I was confined in a cell of my own construction, drained of all enthusiasm. I appreciated every lesson, opportunity, and door opened for me because of the trucking industry, but I longed to build my dream business based on my innate strengths and aligned with my core values.

In a moment of reflection, I chose to seek additional funding, hoping to gain the time I needed to reassess my situation and plan my future moves. I had created a beautiful business that I had nurtured and developed, but it was taking a toll on me. I pondered the implications of letting it all slip away and contemplated the question that lingered in my mind: if I weren't immersed in this endeavor, what path would I pursue instead? The answer was clear; it was a realization I had carried for over a decade. Sometimes, the most beneficial decision is to strip everything away and start fresh, even if it means returning to the beginning.

On a gorgeous, bright afternoon in New York City, the air buzzed with excitement as the graduation practice for Cohort 20 of the Goldman Sachs 10,000 Small Businesses Program unfolded. This initiative, a significant investment by Goldman Sachs, aimed to empower entrepreneurs, enabling them to expand their ventures and contribute to the economic prosperity of their communities. Alongside the educational segment I had been part of, the occasion also offered an incredible opportunity to connect with others in the business world.

As the graduation rehearsal started, I thought about how many business owners also grappled with their own feelings of impostor syndrome. In the days that preceded this significant moment, I often felt stifled by a looming anxiety about the responsibilities and decisions that awaited my return to Louisiana. But instead, somehow, I shifted to consciously envisioning myself striding confidently across the stage, embracing the moment, and embodying my success story. No matter the market fluctuations that had transpired in the months leading up to this day, I had achieved what many deemed impossible, created jobs for dozens, and made a meaningful impact on my community. I steadfastly resolved that self-doubt and insecurity would not overshadow the significance of my achievements.

Transforming my narrative wasn't about donning a mask or feigning confidence I didn't possess. Instead, it involved accepting that my success was multifaceted, and I held the reins to my own story. By internalizing this concept, I decided to exude confidence, confront the assumptions I had formed, and analyze all the information before me as valuable data points rather than merely focus on the unappealing aspects.

As I reflected on my journey through the world of business over the last 15 years, I discovered a framework that transformed my perspective. The 16 business principles of *Mindset Mechanics* became a guiding light for me. I meticulously documented principles that shaped the narratives I constructed about success and my dedication to serving others.

As I contemplated the future, I took a moment to reflect on the people I wished to help and the connections I wanted to forge. With a renewed sense of purpose, I seized fresh opportunities and breathed new life into my business. Starting with a blank canvas, liberated from the weight of previous experiences, I adopted a transformed mindset. I determined my limits, and after I recognized that I had hit rock bottom and felt a sense of sufficiency, I opened myself up to the possibility of learning once more. Over the next ten months, I meticulously crafted a plan and diligently worked towards realizing my vision.

Reimagining the future, I took a moment to reflect on the people I wished to help and the connections I wanted to forge. With a renewed sense of purpose, I seized fresh opportunities and breathed new life into my business. Starting with a blank canvas, liberated from the weight of previous experiences, I adopted a transformed mindset. I determined my limits, and after I recognized that I had hit rock bottom and felt a sense of sufficiency, I opened myself up to the possibility of learning once more.

Over the next ten months, I meticulously crafted a plan and diligently worked towards realizing my vision.

If you don't change your mindset, you risk carrying your old thinking paradigms into your next venture and getting the same old problems. You cannot place new wine in old wineskins. They'd burst.

To avoid repeating the same mistakes, refresh your mind and relearn everything you thought you knew about business. Over the next 10 months, I realized the value of my engineering education. Completing my engineering license was on my bucket list. So, I started studying for the second portion of my Professional Engineering License, and I began to see physics, thermodynamics, and energy in a different light.

- You cannot build a high-performance business with low-quality energy.
- You must find a higher energy source to create a thriving business.
- Energy quality is essential to designing a high-performance business.

I noticed physics and engineering everywhere. To achieve success and growth, I needed to manage my energy. I noticed physics words used by thought leaders who didn't study science. They helped me realize I no longer wanted to contend with friction. I needed to overcome inertia to build lasting momentum. I needed to understand the resultant net forces acting on my business to move it in the right direction. Then, I stumbled into metaphysics, which provided me with an explanation of quantum theory in a more practical way.

I knew I was on to something. From the start, I needed a mental model designed to show other business owners how physics works for any system.

Once a leader under stress, I did not want my experience to go to waste. So, I took time to understand my experiences from a systems perspective. During my crisis, I needed rest, relief, and relaxation. I needed someone to listen and understand me with awareness, and care was key to my discovery. I needed patience and understanding to know that even though I am intelligent, I didn't know all there is to know about sleep deprivation. Now equipped with my experiences and profession, I vowed to tell my story in hopes of helping other leaders become more aware, take care of themselves, and adopt healthy thinking that protects the assets needed to achieve greatness.

I learned several lessons over the years. When it comes to business, the most pressing issue is that our approach to organizations is fundamentally flawed. The way businesses operate today is stress-prone.

After my crisis, I began to look for alternative methods of conducting my business affairs. I found that plan and outlined it here in *Mindset Mechanics.* We find ourselves overwhelmed, anxious, and weighed down by the demands of our careers. We design work that does not align with our envisioned life. This misalignment creates fractures, breaks, stresses, and strains. The fact is, it doesn't have to be this way. Our work can empower and free us at the same time while still creating the wealth and legacy we desire.

I experienced the significance of self-care on a personal level. It's crucial for you to grasp that there is a smarter, more intuitive approach to conducting business that doesn't feel burdensome and embraces right-brain thinking and left-brain organization.

Mindset Mechanics introduced universal principles to explain how business functions, all framed by the laws of physics. When you employ right-brain thinking, work becomes a breeze, and left-brain thinking, like physics, does all your heavy lifting. I genuinely believe that the strategies offered in *Mindset Mechanics* could transform the way you run your business from now on.

In the following chapters, I will give you everything you need to apply *Mindset Mechanics* to your business. When you're finished, I will explain why I've decided to give my entire method away. As for *Mindset Mechanics,* some of the concepts are technical, but stay with me. I'll walk you through every step. The question is: *Are you ready?*

PART II: Applying the Mechanics

SECTION I: The Right Business Vehicle

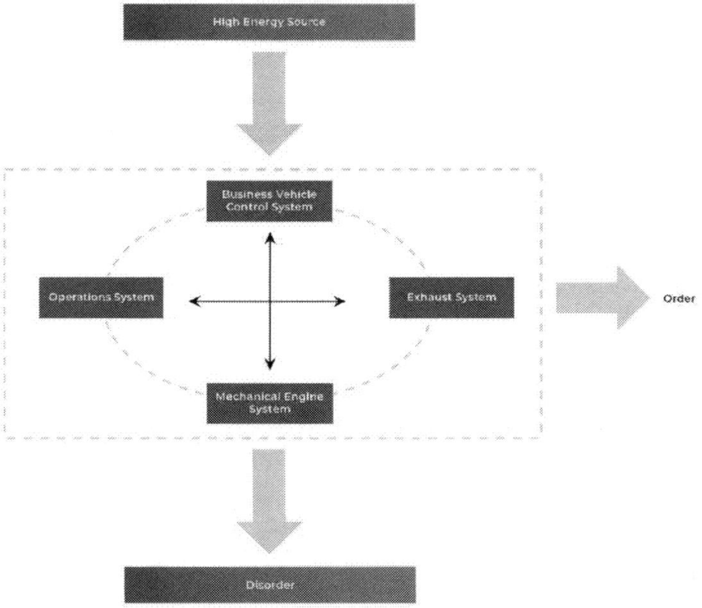

Figure 2. Mindset Mechanics Interface: Business Vehicle Control System

CHAPTER 7. The Core of Success

Imagine your business as a high-performance vehicle built to get you to your desired destination. But like any powerful machine, it requires skillful handling, regular maintenance, and a clear understanding of how its parts work together. Just as a mechanic knows every component of a car, you must understand the inner workings of your business. This is the essence of *Mindset Mechanics*—to help you become a better leader by mastering your business operations and achieving lasting success. This section is dedicated to answering the question: "*How do you drive success for high-performance, happiness, and vitality?*" There's really only one thing that drives success, satisfaction, and vitality, and it's found in the *Business Vehicle Control System (BVCS)*, which includes four main components:

- **The control center of mass** consists of personal and organizational culture: the cultural hub, key people, processes, and structure.
- **The mechanical operator** is the key person of influence, acting as the driver of the business vehicle.
- **The operator's mindset** is the mental framework that guides the leader's decisions, behaviors, and actions.
- **The business vehicle** represents the organization itself, driven by a clear purpose and aligned with your vision.

To ensure clarity, let's define some key terms used in *Mindset Mechanics*, as I may use them interchangeably. An *organizational leader* is the driver of the business vehicle. The *mechanical operator* is a key person of influence who not only leads but actively steers the organization. They access the cultural hub, key people, processes, and strategies to guide the business vehicle. The control center of mass is best developed when the mindset is aligned with three tools:

1. The *Rules of Mindset Mechanics* (rules of the road)
2. The *Mechanical Performance Formula* (manufacturer's operator manual)
3. The *Road to Success* (roadmap)

By understanding these components, you'll be equipped to

operate your business vehicle effectively, align it with your goals, and drive it toward sustainable success. Let's dive deeper into the components of the Business Vehicle Control System, as shown in **Figure 3** below.

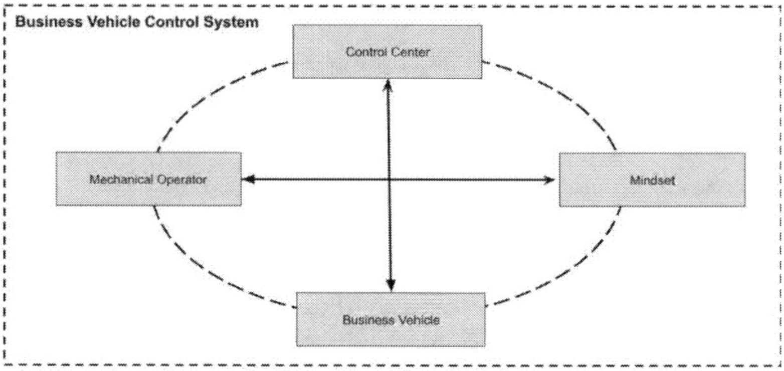

Figure 3. Business Vehicle Control System

Control Center of Mass

The control center of mass is the heart of your business, encompassing personal and organizational culture (the cultural hub), hardware (structure), software (processes), and key people.

Cultural Hub

The *cultural hub* is home to the vision, values, goals, mission, interests, and style that help drive organizational decision-making. I call the process of developing your vision *visionering*. Visioneering is a formula I use to help clients discover each component of their cultural hub to unlock true success, happiness, and vitality. Like most modern vehicles, the control center of mass adapts to the driver's behavior.

Hardware (Structure)

Hardware refers to your team's organizational structure, defining each passenger's role, including the driver, in your business vehicle. It encompasses the mechanical components and forces that power your organization. Key features within the cultural hub help shape the style and function of the organizational structure. Typically, the owner selects a business vehicle based on its hardware, carefully choosing the components that best align with their vision and goals.

Software (Processes)

Software represents organizational thinking, starting with the leader's mindset and adopted by the team. It controls the electrical and power-generating components, as well as the processes and decision-making within the business vehicle. Owners typically choose a business vehicle based on speed, quality, and performance, aiming for the best in class before making a final selection.

Driver (Key People)

In *Mindset Mechanics*, the business owner is both the driver and mechanical operator of their business vehicle (organization). They manage the wear, tear, cleanliness, comfort, and overall experience. The driver is fully accountable for the control center, which helps manage the two main functions of the business vehicle: hardware and software.

The driver is a leader, guiding passengers safely to their destination. But not all drivers are true leaders. A valuable driver plans the trip ahead, knows the route, considers alternatives, anticipates hazards, and makes necessary adjustments. They are prepared for emergencies and can handle unexpected challenges.

The best drivers are mechanical operators—they can detect signs of failure before they happen, ensuring the journey stays on course.

Mechanical Operator

The *Mechanical Operator* is a "key person of influence." Becoming a key person of influence brings greater income, more time, a better quality of life, and increased recognition. There are two types of mechanical operators: *vital* and *functional*.

- A *vital operator* knows where they're going and how to get there, striving to master their craft.

- A *functional operator* simply fills a role and is easily replaced.

The vital operator understands how to read the machine's gauges, interpret functionality, and shift gears to maintain performance. They know how the machine works and can anticipate the effects of every decision they make. Good mechanical operators are not only skilled drivers, but also adept at handling the vehicle.

Rules of Mindset Mechanics

The *Rules of Mindset Mechanics* are the rules of the road. They offer the psychological framework needed to think and respond like a skilled driver and mechanical operator.

Business Vehicle

The *business vehicle* represents the organization, and not all business vehicles are created equal. It's vital to know your destination and terrain before choosing one. You need a business vehicle that can accelerate, when necessary, with a strong engine to handle its weight and the load you'll carry. It should be comfortable, functional, and stylish, ensuring a pleasant experience for both you and your passengers. Aesthetics matters—other drivers will admire the beauty, design, functionality, and appeal of your business vehicle, sparking their desire for one like it. When maintenance is needed, you should have access to top-quality mechanics and parts to keep you on the *Road to Success*.

Every vehicle comes with a manufacturer's operating manual, and *Mindset Mechanics* is no different. Your manual is the *Mechanical Performance Formula*.

Mechanical Performance Formula

The *Mechanical Performance Formula* is your manufacturer's operating manual. It re-centers you when you've lost your way and is the foundation of *Mindset Mechanics*. This formula offers a universal, comprehensive solution for growth.

Roadmap to Success

The *Road to Success* is an indicator of how you view business. It provides landmarks to expect along the way that indicate whether or not you're on the right road. Some see the road to success as rough and rugged, like a battleground, while others see it as a smooth, adventurous journey that feels like a fun game. Your perspective—along with your skills as a driver, abilities as a mechanical operator, and the functions of your control center and business vehicle—determines how well you navigate and enjoy the trip.

It's also crucial to ensure your business vehicle is built for the Road to Success and that you're operating it as intended.

Now that you understand the key concepts of *Mindset Mechanics*, let's dive deeper. The first core function is enhancing your skills to become a Mechanical Operator who can navigate, monitor, and lead effectively.

CHAPTER 8. The Mechanical Operator

Success is more than just a goal—it's a journey shaped by your mindset and how you think and approach challenges. If you've ever wondered why some leaders seem to navigate obstacles and achieve lasting results effortlessly while others don't, this chapter explores the mental framework that drives some Mechanical Operators (see **Figure 4**) forward to achieve success, happiness, and vitality and how this fundamentally impacts how you operate, adapt, and ultimately perform in business.

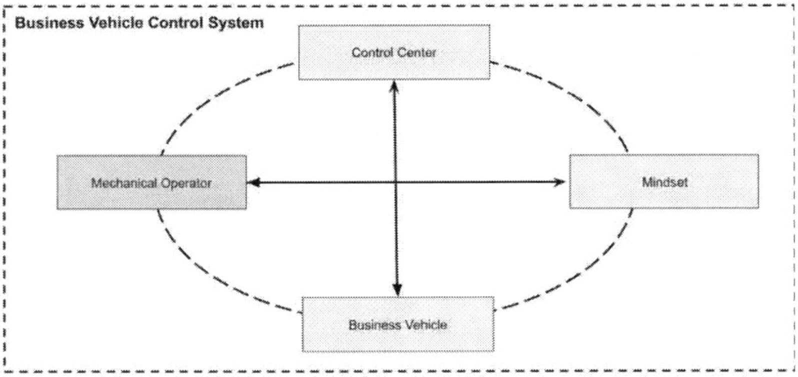

Figure 4. Business Vehicle Control System: Mechanical Operator

So, what truly determines success? According to author T. Harv Eker in *Secrets of a Millionaire Mind: Mastering the Inner Game of Wealth*, success begins with mindset. Eker emphasizes the power of creativity and an intuitive mindset as the foundation for achieving wealth. I would take it a step further: the mindset of the Mechanical Operator (leader) determines success.

This mindset does more than drive ambition. It simplifies complex tasks, connects with the business machine, and anticipates road conditions. It allows for efficient decision-making, fuels management styles, makes strategic choices, and builds effective teams. The leader's mindset is key to reaching goals, saving time, and avoiding unnecessary complexity.

Bell Curve

A bell curve or standard distribution curve is shown in **Figure 5** below. It shows that if you measure any population set, the results will break out with about 96 percent in the middle, and then you'll have outliers of around two to three percent on either end. So, if we were to consider a hundred super-talented students and put them into a standard distribution curve, we'd end up with two to three who are the best of the best.

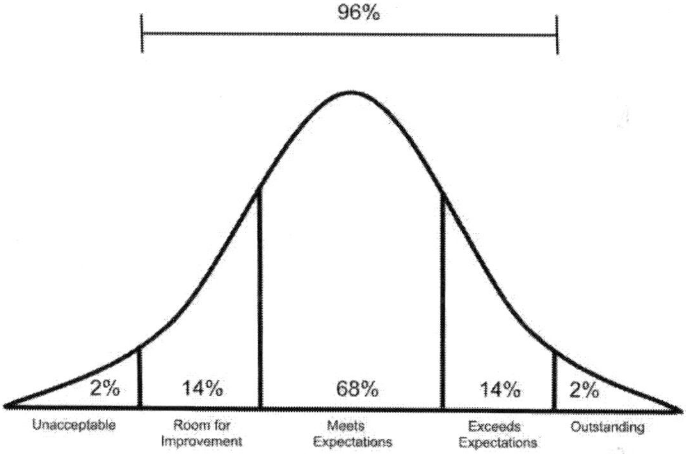

Figure 5. Bell Curve of High Achievers (2%)

Dr. Vance Ceaser explains in his book, *The High Achiever's Guide to Happiness*, that most of us live our lives thinking if we can get into the top 2 percent of the bell curve zone for our area of expertise, then we'd find success and happiness. The data says differently. Dr. Ceaser studied high achievers from all walks of life: politicians, entrepreneurs, artists, and entertainers. He measured their authentic levels of happiness and found only about 10 percent of the highest achievers were also happy. That's 10 percent of 2 percent. So, only 0.2 percent of the population set were high achievers and highly happy. In her book, Dr. Mary E. Anderson calls them *The Happy High Achievers* (see **Figure 6**).

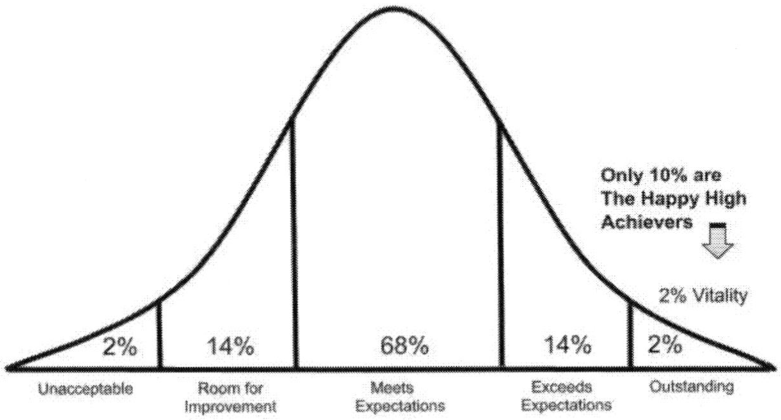

Figure 6. Bell Curve of The Happy High Achievers (0.2%)

So, imagine that you take 1,000 of the wealthiest and most famous people and put them in a ballroom. Only two to three people in that entire room are going to be not only successful but also experience vitality and happiness. What does this mean for you?

Two Types of Mechanical Operators

In *Mindset Mechanics*, an organizational leader is seen as the "key person of influence," acting as the mechanical operator. There are two types of mechanical operators: vital and functional. As Kevin Harrington and Daniel Priestley highlighted in *Key Person of Influence*, becoming a vital leader is more impactful than simply being a functional one. Vital leaders are happy, energized, and successful.

Two types of mechanical operators are also attempting to travel the Road to Success. Anyone steering a business must recognize whether they are a functional or vital leader. Let's explore the roles of these two types of leaders and how they drive organizational success.

Functional Mechanical Operators

Consider a functional person like the moon. The moon's role is to act as a satellite, maintaining standards and executing processes, but it doesn't generate energy. Similarly, a functional person depends on a vital person—the sun—for energy, direction, creativity, and productivity. Functional mechanical operators are not happy, successful, or vital (see **Figure 7**).

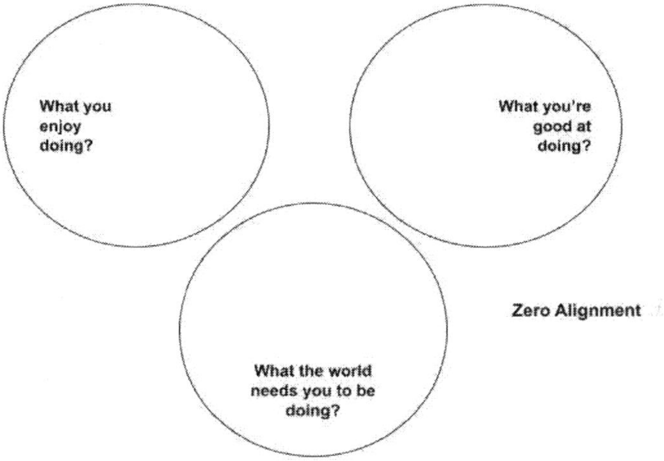

Figure 7. Functional Mechanical Operator: Unhappiness and De-energized.

Functional mechanical operators (see **Figure 5**) often make changes without fully understanding the business vehicle's issues, leading to wasted energy, time, and money. Missteps result in unforeseen problems, making repairs more difficult and frustrating. Without a more capable operator, the business may never fully recover.

The functional mechanical operator lacks alignment of what they enjoy, what they excel most, and what the world needs from them. Functional operators lack of alignment shows up in the quality of their work, outdated skills, and inexperience. Alignment is essential for becoming purpose-driven and leads to greater fulfillment and effectiveness. *Mindset Mechanics* helps you align your purpose, energy, and actions for optimal performance.

Vital Mechanical Operators

A vital person is like the sun in their solar system, drawing energy from within through nuclear fusion. An organization's vital leader isn't energized by teams, employees, or clients but by their environment, peers, and internal drives.

Like the sun, when a key person of influence is absent, people notice and anticipate their return. Harrington and Priestley describe this vital person as having the "life-force" — the energy that fuels others. They are aligned and keenly aware of their passion, purpose, and position in life and within their organizations (see **Figure 8**). Vital leaders know their role is more creative than functional. Managing energy starts with managing health — getting rest, recharging, and staying prepared for their passion-driven work.

Vital leaders know their role is more creative than functional. Managing energy starts with managing health — getting rest, recharging, and staying prepared for their passion-driven work. Burnout affects not just them but the entire system they lead. To prevent it, a vital mechanical operator must align what they enjoy doing, what they excel at doing, and what the world needs most.

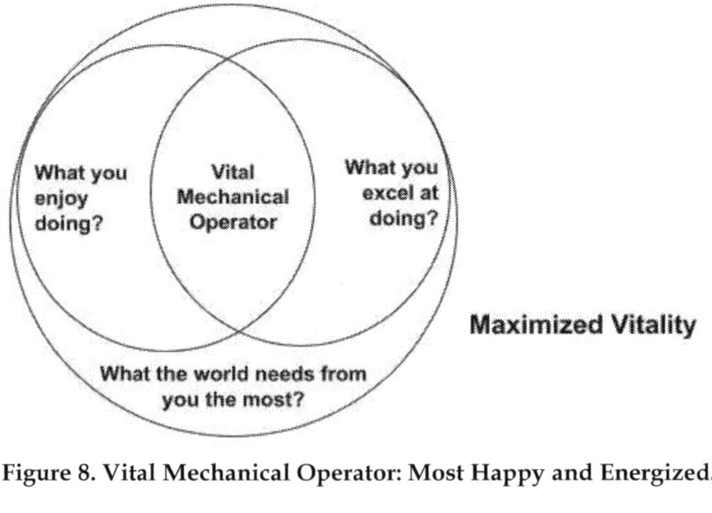

Figure 8. Vital Mechanical Operator: Most Happy and Energized.

Are you a vital or functional leader? As a mechanical operator, you must choose to be either vital or functional. Vital mechanical operators prioritize alignment, continuous improvement, and good maintenance practices. They enjoy their work, rely on expert networks, and understand the importance of sharing progress through documentation. They maximize their vitality by aligning their strengths, others' needs, and personal fulfillment.

Battle vs Game

The Road to Success guides your business, but road conditions impact both you and your team. Imagine driving on a bumpy road—it's slow, potentially damaging, and exhausting. After such a rough journey, you arrive worn out. Now ask yourself: Are you following the right roadmap? When you look back, do you like what you see?

98% of functional mechanical operators drive the road to success and experience a battleground.

Most leaders—98%, are functional mechanical operators, viewing business as a battle and the market as a battleground. But what if there was a better path? You can transform your journey by shifting your mindset, upgrading your business vehicle, enhancing your skills, and following a clear roadmap. Instead of conflict, think of business as a game with winners and losers. But what sets them apart is what makes the difference.

Most leaders fail because they focus on the wrong things. They invest in equipment, education, and training but struggle to translate these resources into meaningful results. They add titles and certifications to their names but remain stuck in a reactive state rather than being responsive to their environment.

Winning in business starts at the individual level by responding to mass, inertia, and momentum and not reacting to them. Becoming more responsive separates you from the 98% who simply react to perceived threats to their value. Choose the right road and change the game for yourself and your organization.

2% of vital mechanical operators drive the road to success and experience an enjoyable game.

The 2% of successful mechanical operators view business as a game and treat it like their playground. They know the road to success is built on preparation, performance, and transformation. These leaders focus on who they're becoming, their destination, vitality, and sustaining a growth mindset.

They control their game using invisible forces—monitored through gauges in their control center—that ensure comfort and track performance. Vital mechanical operators understand how to apply force and steer their business vehicles effectively, adjusting speed and direction based on conditions, weight, and loads. They know how to navigate toward success with precision and purpose.

How Vital Leaders Win the Business Game

How can you accelerate your business to win the game? You can't do it with just any old equipment. The road, skills, team, and business vehicle must fit the job and allow for speedy implementation and improvement. Otherwise, the journey will only get bumpier and filled with more frustration. Attempting business without understanding these critical success factors becomes a failure to prepare, leading to a rough ride where your only focus is survival, not innovation.

Let's face it, most organizations have no idea what I am speaking of. They've built their businesses in isolation, isolating them from the desired success. As a result, they're driving down the wrong road with a poorly equipped business vehicle and the wrong focus. It makes sense why they feel limited time, energy, resources, and attention pressures, making momentum hard to build. The functional mechanical operators struggle to accelerate their businesses because winning requires good decision-making, speed of implementation, and intentional change. No matter how hard they try, how much money they throw at their challenges, or how talented their team is, nothing seems to work, and they don't have the fundamentals to accomplish their goals. You need to understand how change affects you, which principles to apply, and how to access the proper roadmap—or risk speeding toward failure. They'd be better off driving blind.

98% of leaders fail to be intentional, allowing their business vehicles to suffer in battle and lose the game. But for the 2% of vital mechanical operators are strategic about success. They play to win by becoming key influencers, following operational guidelines, maintaining performance, and mastering the rules of the road. You can join the 2% and accelerate toward success with the right approach.

Winning the game of business is not about working harder but thinking smarter.

Suppose you want to join the 2% of vital mechanical operators, win the game of business, travel the road to success, and experience happiness. In that case, you must understand that success is rooted in your vitality. You can't win if you're unhappy or exhausted.

Be Fulfilled

You can't operate from an empty cup to win the game of business. You need strength, resilience, and a love for your work to push through the inevitable—challenges and tough decisions.

Let's do an exercise before you begin making changes. Please identify the tasks, people, and situations that drain you most. Are you exhausted because you lack the right tools, skills, or knowledge? Are you expected to be someone you're not? Do clients fail to appreciate the work you do? Do employees see you as an adversary or an ally?

For the 2% of vital mechanical operators, work is fulfilling and fuels them through tough times because they surround themselves with things they love, and the people they work with and serve excite them. For them, their work is their life's mission. But for the 98% of functional operators, work feels unbearable because they lack internal motivation.

For them, their work is a means to an end.

Whatever you dislike doing for long periods, create a plan to stop doing it eventually. When you started your business, you may have to start by wearing many hats, but as you grow and learn more about yourself, you should pay attention to how your body and mind react to your work. Then actively look for your replacement—not just any warm body will do. Look for someone who loves and finds fulfillment in doing the tasks that drain you. This means you must be intentional.

You are the vital sun in your solar system. Operate from a place of fulfillment—because you can't give from an empty cup.

Be Energized

As a vital leader, you are the sun in your solar system. But even the sun needs a renewable energy source to continue fueling others. What do you enjoy? What keeps you going? As you pay attention to what energizes and drains, you create rituals that renew you. To be a vital mechanical operator, you must align with people, ideas, and activities that energize you. Energy and its sources are found outside your organization, in your environment.

98% of functional operators draw energy from within their system—teams, clients, and vendors. But these sources eventually deplete, leaving the organization drained. According to physics, *energy must come from external sources at higher levels.* You can't solve problems, innovate, manage, and serve from the same or lower energy level.

Another thing is that the 2% of vital mechanical operators don't depend solely on their strength or their team's—your vitality is not your team's burden. I believe being a strong person on its own is overrated and exhausting. Instead of being strong, rely on building your courage and faith, then let universal forces do the heavy lifting.

As the energy source of your solar system, your vitality and energy drive everything. Prioritize energy management to keep your organization alive and thriving.

Find Alignment

You must understand that the game is played with invisible forces to win. It starts with aligning your source of fulfillment and energy with these invisible forces. These forces are an inner drive and self-worth that guides all principles. When aligned, your happiness and energy become invaluable—others want to be around you. People call it charisma, but I think it's much more. Finding alignment develops into an inner light force that attracts others, elevates your quality, and shapes your focus. They draw people to your vision, mission, and organization. You move and operate with a sense of belonging in the world, which cannot be explained but can be felt and observed by others.

The 2% of vital mechanical operators succeed because they have found alignment. They understand that alignment allows them to focus solely on their core desires, which fuel their vitality and bring unlimited happiness and energy. Alignment is the force that makes all the difference.

How to Build Alignment

A vital mechanical operator focuses on aligning their inputs to get desirable outcomes. The first step is to focus on building your internal culture. Internal culture is your vision, values, goals, mission, interests, and style. Your inner culture is the input for your organization's culture, which creates the structure and processes needed for business building. Interestingly, your organization's culture mirrors your inner culture, so this step is important. If your business is chaotic, your inner world is most likely experiencing chaos. If your team is confused about what they are doing, you and your process are most likely unclear. I think you get it. The key to becoming a vital mechanical operator and building alignment is three questions (see **Figure 9**).

1. **What You Enjoy Doing:** How do you prefer to spend your time and energy?
2. **What You're Good at Doing:** How do you spend your time and energy?
3. **What the World Needs You to be Doing:** The demands and

expectations others place on you.

Answering these questions prepares you to refine your internal culture, structure, and process for managing your organization. Now, let's define your inner culture as a leader—the culture hub inputs into your vision model, V*isioneering,* to engineer your vision.

Figure 9. Aligned Vital Mechanical Operator (VMO)

Internal Culture Conflict Resolution

As a leader, you must build your inner culture, which is a key component of your center of mass based on your vision, values, goals, mission, interests, and style that's aligned and conflict-free. **Vision** guides you toward your goals, while values are non-negotiable principles. Most functional operators (98%) believe they must sacrifice one area of life for another, but true success requires alignment with what you love, where you're headed, and your business offerings. Don't compromise. If you're not excited about what you do, most likely, that's a clue that there's a conflict in your internal culture and you've compromised, which will not bring success, happiness, or fulfillment. Vitality comes from focusing on what truly aligns with your core values, purpose, big dream, and mission.

Your core values should come from getting to know yourself—they define who you are and should rarely change. Your purpose is the fundamental reason your company exists, explaining the value you bring to the world. Your big dream is a bold, measurable, and inspiring goal, while your mission is clear, concise, and achievable.

Your goals are your ambitions and efforts to achieve the desired result. When conflicts of goals arise, two or more goals compete for resources that cannot be achieved simultaneously. Recognizing and resolving a conflict of goals is one of the most important tasks in aligning an organization and mobilizing a team on a united front. Conflicts of goals arise for several reasons. Maybe resources are scarce, goals are interdependent, or goals are incompatible. As a vital leader seeking alignment, you must choose your primary goals based on your desired outcome and order of importance.

Sometimes, a conflict of mission exists within your inner culture due to a lack of clarity. The mission is your objective and actions to reach the desired outcomes. When there is a conflict in the mission, there is a lack of alignment, clarity, engagement, or focus. The team does have a clear sense of what to do, the effects of each action, and the desired outcomes.

The next most significant source of mass is a conflict of interest. A conflict of interest means that what's good for today isn't necessarily good for your tomorrow. A conflict of interest can be negotiated, and you do that by setting parameters and aligning your inner vision and values.

Conflict of style means that your management and behavioral styles might differ radically. Suppose you're a highly innovative, creative individual. In that case, you see every problem as an opportunity to develop a new solution. Still, when you're around family, you're a more detail-oriented person who prefers to move more methodically and prudently, and maybe you like to control things. Being aware of finding alignment is important to maintaining vitality and finding happiness. Your core culture must be congruent on all fronts. Your style might rub others the wrong way and vice versa. The language you use, the pace at which you move, the approach you take—all those things can create a conflict of styles. You must seek authenticity and personal development in style.

Now, just like a conflict of interest. A conflict of styles can be harnessed in such a way that it leads to better decisions that get implemented more quickly. I'm going to show you how to do that. So, these are the three sources of mass in an organization, and two of them are negotiable; they can be harnessed so that conflict becomes constructive.

A culture of internal conflict is something you cannot allow in the organization. So, it would be best if you fixed it within yourself first. If there is unresolved internal conflict within the leader, it could lead to a split of the team or a disconnect in the organizational performance.

The secret we'll discuss in Chapter 24 is that your inner culture should align with what you do, how you interact with others, structure your life, and make decisions. This alignment is the foundation of a high-performance culture for individuals and organizations.

Inner Person & Inner Tribe

Who do you want to be? How do you see yourself, and how do you want others to see you? Building vitality means aligning your inner vision with how you present yourself to the world, which starts with your health. True exterior health reflects interior well-being. Knowing who you are, what you're becoming, and your weaknesses is essential to finding alignment and vitality. I want you to know that personal deficiencies aren't negatives; they create opportunities for others to contribute their strengths. They are opportunities for everyone to become vital.

As the sun in your solar system, you must generate the internal force that keeps you and your business in motion. Who energizes and motivates you? Who do you need to become to add value to them? Surround yourself with people who enrich your life, creating a mutually beneficial relationship.

Seth Godin defines a tribe as "a group of people connected to one another, connected to a leader, and connected to an idea." When you clear your vision and values, you'll attract the right people actively seeking you and your company.

Internal Structure is Design and Function

Structure is about design and functionality. What kind of business vehicle will you create? Does the design fit the road (market) you want to travel (serve)? Who will your business serve, and who are the passengers (teams) that will support you? Be intentional about how you structure your business.

Design your business to align with the lifestyle you want. Guess what? You get to decide the size of your business, your work schedule, and your role as the leader. Do you want a large or small business? Will you work a 4-day workweek or 4-hour days? Will you build a flat or hierarchical team? The choice is yours—do you want to be a functional or vital leader?

Functionality is about purpose. Build a structure that serves a clear purpose. Define your role and determine how many employees you need. Structure adds weight, affecting speed and decision-making, so design it to fit the market and serve your target audience.

Vital mechanical operators know the value of their time, so they design businesses that foster autonomy, enabling others to share in decision-making. Structure shapes your business vehicle—be intentional about the ideas, materials, and reasons behind how you design it to function.

Internal Processes Go Slow to Go Fast

As a vital mechanical operator, you've aligned your internal culture—refining your vision, defining your values, and structuring your lifestyle and business. Now, it's time to develop your process, which has two parts: decision-making and implementation. Here's where you must *go slow to go fast*.

In business, decision-making and implementation have a rhythm. Many want instant results, but that's not the same as quantum leaping.

The first skips crucial steps; the latter strategically streamlines outcomes to accelerate progress.

Go slow to go fast means spending quality time making decisions and then traveling at the speed of light to implement them. Most get it wrong, are hasty to make decisions, and slowly implement them. *Go slow to go fast* means taking quality time to make decisions, then implementing at lightning speed. Most do the opposite—they rush to make decisions and drag to implement.

In *The Slight Edge*, Jeff Olson emphasizes the power of time in due process: you plant, cultivate, and harvest. Time magnifies consistent, small actions, connecting your efforts to a multiplier. Your process is the seed, and the laws of physics provide the time to yield results.

As a vital mechanical operator, align your internal culture, attract fulfillment, and apply the laws of physics—like Newton's Law of Gravitation—to achieve your desired outcomes.

Newton's Universal Law of Gravitation

Vitality is your ability to *align* your desires with the universal laws of physics and then apply organizational principles to build and sustain *momentum*—an essential measure of motion. If you're stuck or moving in the wrong direction, inertia is likely the cause. Inertia halts progress and prevents a change in direction. Aligning your internal culture with the laws of physics transforms desires into momentum.

However, your deepest desires—whether to achieve wealth, escape financial hardship, or gain more leisure—must not be superficial. These are outcomes, not drivers. Your core desire must be rooted in serving others, not purely for altruism but to find a purpose larger than yourself. This greater cause provides the 2% leverage needed to set momentum in motion and overcome obstacles.

Before diving into momentum, let's consider gravity's role in motion. Isaac Newton discovered that gravitational pull diminishes with distance but is also tied to mass. *Newton's Law of Universal Gravitation* (see **Figure 10**) states there exists a force of attraction between two objects (m_1 is the internal mass of the system, and m_2 is the internal mass of the desired object or outcome), and this force is proportional to their masses and inversely proportional to the square of the distance, r^2, between them.

$$F = G\frac{m_1 m_2}{r^2}$$

Figure 10. Universal Law of Gravitation Formula

Understanding this universal principle of attraction allows you to harness the forces that drive your business forward, aligning with the laws of physics to generate powerful momentum. But how does this relate to your organization? I'm glad you asked. Let's define a few terms.

Mass Attraction and Gravitation

Newton discovered that everything with mass has a proportional gravitational attraction. Even as a human being, you exhibit forces of attraction that are influenced by your mass (personality or mental attitude), proximity, and vibrational frequency to other masses (objects or people with personality and mental attitude).

Mass is not the same as weight. Mass measures an object's internal matter and remains constant, while weight is a force that acts as a downward pull to help objects stay grounded, which could change. Mass focuses on the amount of matter in a given space. Matter is the substance that makes things—living and nonliving.

All matter resonates at a specific frequency—whether it's a chair, car, or house. A chair resonates at this frequency. A car resonates at the car frequency; the same goes for a house. Everything that is made or created begins as an idea that merges the mental (invisible matter) and physical (visible matter) into useful objects. Two objects of different masses will attract based on their distance and vibration. When two mass frequencies are equal, they naturally gravitate toward each other with the attraction force—just like the saying "birds of a feather flock together."

To attract what you desire—whether it's employees, resources, or outcomes—you must first become what you desire and then vibrate at that same frequency. Over time, your organization will naturally align with and attract what you need to succeed.

Start with Attraction and Mass Accumulation

In our universe, everything vibrates and possesses mass—including you, as you engage in the delicate dance of vibration, attraction, and even repulsion. Leaders often unintentionally attract undesired outcomes due to a lack of understanding of these universal forces, resulting in frustration and confusion.

Everything in your life exists as invisible matter, shaped by your conscious or unconscious participation—it becomes physical matter. Whether intentional, solicited, or not. Nothing is by chance—everything has a purpose that you've created. You have the power to shape your reality. Creating with awareness, alignment, and intention is much better.

Don't leave things to chance—embrace your power and align with universal principles. Successful leaders understand the role of attraction. You can steer your journey toward your desired outcomes by intentionally accumulating and attracting value.

Attraction Builds Mass

The journey to your desired reality becomes clear once you understand and apply physics principles. I want to make a clear distinction. This isn't about the "law of attraction" you may have learned online—I've never encountered it in my engineering studies, nor has it worked for me. Instead, we will focus on what works based on the laws of physics.

To achieve your goals, envision your future self, align your actions, and engage in reflective practices. Let your thoughts and conversations mirror this identity, and ensure your actions resonate with your vision.

This alignment is the key to transformative change.

The Butterfly Effect

Building mass is similar to the Butterfly Effect in chaos theory, coined by mathematician Edward Lorenz. It illustrates how small changes can create significant future impacts—by making one choice, you can transform your entire life.

Consider the journey of a butterfly. As a caterpillar, its initial goal is to build mass, feeding voraciously to prepare for metamorphosis. It sheds its skin during this stage, growing rapidly until it reaches the pupal stage. Here, its accumulated mass fuels a complete internal transformation. Dormant cells activate as they reach their potential, shaping the butterfly's wings, legs, and eyes.

Once transformed, the butterfly emerges, vibrant and free, ready to find a mate and sustain its lineage. If it fails to recognize its new state—remaining in a caterpillar mindset—it misses its purpose, preventing future generations from experiencing its glorious beauty. Like the butterfly, you must shed your past, exploit your potential, and fully embrace your desired reality. Notice how small changes in your desires, language, habits, and thinking can have enormous effects on your future outcomes. Amazingly, these changes compound over time as you attract the future leader and life you want and deserve.

Finally, at the core of attraction and transformation is the law of giving and receiving. Allow yourself to receive abundance, then share it with others. To build your inner culture as a leader, align with the Law of Universal Gravitational Attraction—resonate with the frequency you desire to have in your life and allow these forces to attract you to the future person you aspire to become.

---

16. Harrington, Kevin, and Daniel Priestley. *Key Person of Influence: The Five-Step Method to Become One of the Most Highly Valued and Highly Paid People in Your Industry*. Rethink Press, Limited, 2015.
17. Allen, James (1901). *From Poverty to Power*, p. 15-22.
18. Rohn, E. James. *The Seasons of Life*. Jim Rohn International, 1981.

19. "Butterfly Life Cycle," September 14, 2020. https://ansp.org/exhibits/online-exhibits/butterflies/lifecycle/.
20. Lorenz, Edward N. "Deterministic Nonperiodic Flow." *Journal of the Atmospheric Sciences*, vol. 20, no. 2, 1963, pp. 130-141. *Deterministic Nonperiodic Flow*, https://journals.ametsoc.org/view/journals/atsc/20/2/1520-0469_1963_020_0130_dnf_2_0_co_2.xml?tab_body=pdf

CHAPTER 9. The Control Center

Designing your life and business is much like customizing a vehicle. It's an exciting process that requires thoughtful choices, a clear vision, and a focus on creating something that aligns perfectly with your desires. In *Mindset Mechanics*, we explore how this metaphor not only applies to cars but also to building the business of your dreams—one that reflects your values, purpose, and goals, which are found in the *control center* of mass for your organization (see **Figure 11**).

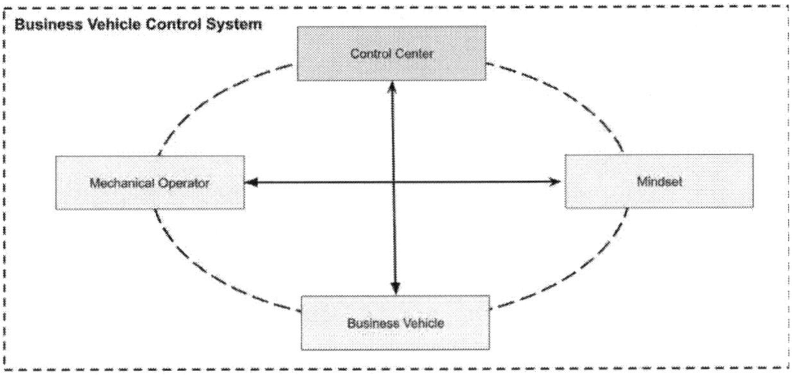

Figure 11. Business Vehicle Control System: Control Center

If you've ever ordered a custom vehicle from the manufacturer, you know it's a fun and immersive experience. You get to choose the make and model, interior and exterior colors, and features—right down to the engine under the hood. With no budget limits, you create your dream car.

But after driving it for a while, you begin to see what you could have done differently. Some features may be less important than you thought, while others need updating or replacement. Once you receive your new vehicle, you try to sustain that fresh look, smell, and feel for as long as possible. As the driver or mechanical operator, you are at the control center of mass. You design and structure the car to your liking, set the rules for passengers, shape the internal culture, and make all the decisions. You determine the intention, energy, and force needed to maintain and operate your vehicle effectively.

Just like a car, your business needs regular adjustments, improvements, and alignment to perform at its best. It's not just about the initial build; it's about the ongoing commitment to refine and enhance it, ensuring it stays aligned with your vision and reaches your desired destination. But remember, the vehicle is the object, while the mechanical operator's mindset is the subject that drives everything. Your business will start to reflect your behaviors—if you care for it, it shows; if you neglect it, well, that shows, too.

As the driver, your mindset and decision-making determine the level of intention and effort you put into maintaining and operating your business. Ultimately, it's your approach that shapes its success.

Organizational Cultural Hub

The core of the control center of mass is the organizational cultural hub—vision, values, mission, goals, interests, and styles. Each of the cultural hub components defines the purpose of your business and its purpose. Culture contains the guiding principles for how your organization responds to the environment and shapes the internal organization.

Whether you're creating an app to help busy mothers manage schedules or designing a juicer for health enthusiasts, culture drives what your business was created for and the value it brings to the world. At the heart of the control center of mass is the cultural hub. This hub is dedicated solely to the organization's culture—its vision, values, processes, and people. (see **Figure 12**).

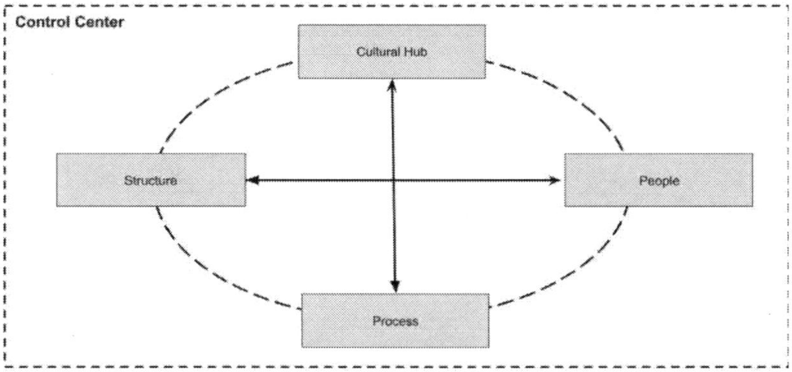

Figure 12. Control Center (Cultural Hub)

Organizational Vision

Vision is a destination. Where do you see yourself in 12 months? Where do you hope for your organization? What will it feel like once you arrive? Who do you need to help you get there?

When describing your vision, speak and act as if it's already here. The more you believe in its reality, the more attractive it becomes. Vision activates Newton's Universal Gravitational Force of Attraction. Your vision shapes how you communicate your destination—making it compelling for others who trust you're heading somewhere worthwhile.

When you share a clear vision that others can see themselves in, the laws of physics work to draw them to your organization as employees, leaders, partners, or vendors. Your ideas and behaviors—rooted in your values—create a magnetic pull, attracting those who align with your mission.

Organizational Values

Your values define what you believe in, what matters to you, and your non-negotiables. When something is important, you take action. To activate Newton's Law of Gravitational Attraction, others must see in you what they see in themselves, or they must see what they aspire to have or become. They need to see your values and what you stand for. Your values need to be visible through your actions and behaviors. It's not enough that they are just beliefs. They must be beliefs in action.

Marva Collins, founder of Westside Preparatory School in Chicago, embodied this by "acting out values," allowing everything she did to reflect her beliefs. As a result, she attracted deserving students, top teachers, media attention, entertainers, and even presidents to her mission for educating underserved children.

Organizational People/Vital Person

Just as you've defined who you will be as the leader, you must also define the people needed within your organization. What types of people motivate you? Who has the motivation and skills to do the work? Who adds value to both the organization and the client? Who can best serve teams? Identify the types of people who bring positive energy and those who drain it. Build a tribe, not just based on skills but also personality, with you as the chief.

Once you attract them, keep them engaged. Ensure they are driven to learn, aware of their strengths and weaknesses, and placed in roles that allow them to develop and express their highest potential. Create opportunities for them to build influence and integrate fully into the organization. Help them shine.

In my experience, people don't quit businesses; they quit leaders who fail to see their worth, provide growth opportunities, or nurture talent. Build an organization that keeps teams challenged and engaged, and you'll create a business not filled with employees but a tribe of raving fans.

Organizational Structure

Just like you defined your lifestyle and function as a leader, you must also define the structure of your team (see **Figure 11**). This is how they interact with each other, their focus areas, and how they interact with the environment. Architect Louis Sullivan famously said, "Form follows function," meaning the design of any building or object should be driven by its purpose. Your business structure should also reflect and align with the cultural hub.

Consider why a car, rocket ship, or boat is designed the way it is—their forms match their intended functions. If you want a business that is agile, recession-proof, establishes authority, creates in-demand products, pays employees well, or scales quickly, its structure must align with its function. Purpose comes first; structure follows.

Organizational Process

As you define your decision-making process (see Figure 11), ensure that it aligns with the cultural hub and organizational structure so your team can make and implement decisions quickly. Determine: *What are your key processes? What do your clients value most? How do you process information and make decisions?* Your decision-making approach should resonate throughout the organization. If there is not a clear decision-making process, your team will second-guess themselves, they will make mistakes, and you will find they continuously have to check in with leadership, not because they are not capable, but because the decisions they need to make have not been clearly defined in processes that drive what they do from day to day.

To build a world-class, high-performance culture, ensure processes are aligned, and decisions are made at the right level. Large businesses can learn from smaller ones about being lean and agile, which enhances decision-making and market responsiveness. A key advantage of small businesses is their speed of implementation—an edge that all organizations should strive for.

Your passengers should be people of value and fit in your cultural hub.

Returning to our analogy, the passengers in your business vehicle are the people who manage, maintain, guide, and drive your organization. They must align with your vision and values, have a defined functional role (even if you have to create one), understand your decision-making process, and be motivated to step into leadership roles. They should also grasp the mechanics of your subsystems and overall operations.

Your passengers must fit within your cultural hub, contributing as "cultural adds," as Danielle Leslie puts it—they are the people who enhance the business and add value.

Business Vehicle Control System

The Cultural Hub comprises the Mechanical Operator with knowledge of the Rules of the Road, the Control Center, passengers aligned with Organizational Values, and the Business Vehicle—all forming the Business Vehicle Control System.

This system is where the operator's mindset and business vehicle connect, creating a cohesive control center. It adapts to reflect the operator's beliefs, experiences, and expectations.

Before diving into the business vehicle, let's first cover the mechanical operator's Rules of the Road for *Mindset Mechanics*.

Rules of the Road

Before hitting the road, conduct a thorough check of the mechanical operator. Run diagnostics on their mindset, ensure proper application of internal and external forces, and confirm they are passionate and driven.

Your mind is like a supercomputer, ready for a software upgrade. It's time to shift from your current mindset to one that aligns with your future self. Before implementing strategies or changes, this is the ideal moment for mechanical operators to adapt to the Rules of the Road to Success—core behaviors and competencies essential for *Mindset Mechanics*. Here are the key rules:

A. Run Internal Systems Check

The mechanical operator's mindset and control center work together like a supercomputer. When systems become outdated, it's time to upgrade both the software and the operator's mindset. Now is the moment to align with the mechanics of your future self by using the *Mindset Mechanics Vehicle Control System*. Start by running diagnostics—change is impossible without awareness.

1. You can't improve what you don't measure. Trust the visionary leader within you and start tracking results.

2. Stevie Wonder's *Hotter Than July* album captures this concept perfectly, especially the song *As If You Read My Mind*. It's about tuning into what you've been neglecting and focusing deliberately on your desires. This journey requires imagination and mindfulness of your current speech and thought patterns.

3. Run internal diagnostics to rediscover your happiness, passion, and drive. What's missing in your life and organization? If it ignites you, it's likely to resonate with others too.

B. Find Internal and External Alignment

Success thinking reflects your mental attitude, aligning directly with your organization's behavior. It comprises four components: culture, structure, process, and people.

- **Culture** includes your organization's vision, values, mission, goals, interests, and management style.
- **Structure** is the support system that organizes people and processes to sustain the business.
- **People** should have diverse personalities, acting as forces that drive resources, information, and success-oriented mindsets.
- **Process** guides decision-making, energy exertion, and idea development.

Each component of your mental attitude plays a crucial role in your business's operation. Before diving into these components, it's essential to establish the Rules of *Mindset Mechanics* around successful thinking.

C. Follow Your Passions

By following the Rules of *Mindset Mechanics*, you have the potential to achieve the success you've always desired. It all begins with envisioning your most ambitious aspirations. I'll reveal strategies used by the top 2% of successful business owners. Now is the time to share your unique truth with the world. Embrace love and unity for humanity, inviting only what aligns with your values. Make it a priority to align every aspect of your life with this love.

This is the essential path to receiving what you truly deserve. Years ago, I picked up a book titled *Do What You Love and the Money Will Follow* and failed to dig deeper. Despite its compelling message, I dismissed it as unrealistic and fell back into old habits. But deep down, I knew it was true. I wanted to change my mindset and break free from old patterns, yet I struggled to know how.

How can you pursue your dreams if they don't generate income? How do you focus on one aspiration amid so many talents? The answer lies in zeroing in on a singular desire—one so powerful that it draws you toward your best outcomes. As Stevie Wonder sang, *"All I do is think about you."* What's that one thing that constantly occupies your mind? Embrace it, and let it guide your journey.

This should be your main priority. For years, I neglected my right to success and happiness. Pursue your passion wholeheartedly. Reflect on what brings you joy, articulate your appreciation, and focus on it. If you do, everything else will follow. I am proof of what's possible, and soon you will experience it too.

Before operating your business vehicle, the mechanical operator must adapt the Rules of the Road by following the *Rules of Mindset Mechanics*. These rules help develop the mindset necessary for driving success, sustainability, and speed toward your goals. You can implement strategies that improve your results with an open mind and willingness to adopt these ideas. Let's start by focusing on developing a mindset that welcomes success.

***"You may bring about that improved condition in your outward life which you desire if you will unswervingly resolve to improve your inner life."* James Allen**

Mindset Mechanics is about mastering your thought process. How do you perceive yourself, your organization, and your future? In the next chapter, you'll find 16 rules designed to help you cultivate the mindset needed to fully embrace and implement *Mindset Mechanics*. These rules will unleash your inner mechanical operator, driving performance, sharpening focus, and creating an internal culture that aligns with and reflects your mental attitude.

Now, let's put into practice running diagnostics, finding alignment, and following your passions. Here are 16 foundational Rules of *Mindset Mechanics* that shaped this book.

CHAPTER 10: The Rules of the Road to Success

Imagine a world where business feels effortless, success flows naturally, and satisfaction is the norm. Sounds ideal, right? The secret lies in the power of the pivot. When used correctly, pivots can reshape your path, opening doors to freedom, success, and fulfillment. They're not just about change. They're about making smarter moves that bring desired results with less effort. Pivots are crucial because they realign you with your goals, setting you on a better track toward success. But what if everything feels like an uphill battle? If each gain has been hard-won and change brings anxiety, you might be fighting for results that don't truly align with what you want—or you've made compromises that still haven't worked out. If business feels like a constant battle, can you truly say you're winning the war? Are you seizing moments of opportunity when things improve? If not, you may need to revisit the *Rules of the Road to Success,* which focus on cultivating a new mindset—one that turns challenges into opportunities and leads you to lasting success. Now, we'll cover the mindset needed for the mechanical operator to manage your organization's *Business Vehicle Control Center* (see **Figure 13**).

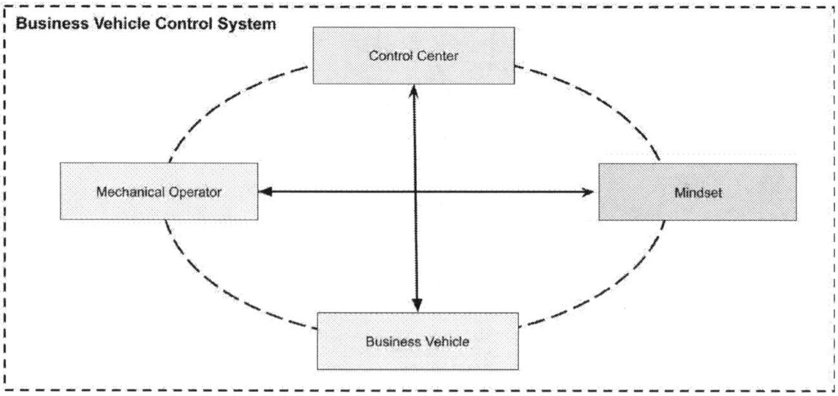

Figure 13. Business Vehicle Control System: Rules of the Road consist of mindset.

In engineering, the term "moments" is also called pivots. No matter how hard you work, failing to "pivot" and seize your "moment" of opportunity can lead to falling short of your goals, misalignment, struggling in business, and experiencing emotional frustration. Pivots are crucial, but *how* you pivot is what truly matters. Before making your next move, consider this: Everything you do and create is a reflection of your inner thoughts—an ongoing cause-and-effect cycle.

British philosopher James Allen called it the "Vision of the Law," where your business reaps what you've sown. If you see your organization as a battleground, it will remain one. Without intentional change, nothing will improve. As Jim Rohn put it in *The Seasons of Life*, your business is a collection of the people and events that have shaped you, reflecting the quality of your relationships and leadership.

Successful pivots happen when you understand the invisible forces working in your favor. To pivot effectively, you must first grasp the *Rules of Mindset Mechanics.* Only then can you create meaningful, lasting change in your business and achieve the success you envision.

Mindset Mechanics Rule #1: Organizational improvement and transformation are mental sports.

The key to organizational improvement and transformation is to be the person who demonstrates change and does the things that drive change, and you'll have your desired output. Unfortunately, 98 percent of leaders believe that by taking action. However, two percent of leaders achieve their desired outcomes by embodying the desired emotions that success brings. You can't transform systems by doing, working, or taking action. You must become the person first.

Desired emotion is a mental attitude, and for a team or organization, it shows up as a collective mental attitude. 98 percent of leaders believe that if they make the right choices, they can achieve their goals and find contentment and success. However, the laws of thermodynamics prove putting in extra effort does not lead to attaining desired outcomes.

Sustained desired results and transformation are about providing each business subsystem with what it needs while at the same time integrating each individual and component and enabling the entire organization to make informed decisions. First, the leader must get what he or she needs by developing a positive mental attitude and vision for the future.

Functional leaders will act to achieve an outcome rather than their goal. If they need to fill an open position, they will attend a hiring event and post the new position online for a couple of months, but when the new hire is brought on board, they often struggle, underperform, or hurt the team because the hiring decision was solely focused on finding a superficial fit.

A leader must stay focused on their organization's goals. The actions of a functional leader may yield immediate results, but ultimately, they find themselves continually seeking temporary solutions. These leaders work tirelessly without realizing that they lack a deep understanding of the real causes.

Once you grasp the true nature of transformation as a mental sport, you will begin to view yourself, your team, and your organization's involvement as something more than just a collective of activities. It's not about the tangible actions you take but rather about the underlying efforts driving your cause. As you read this book, you'll understand and manage these efforts.

Mindset Mechanics Rule #2: You determine the rules in your game.

The concept of truth is simply an idea. You've got your truth, and then there's the ultimate truth. These two must harmonize to effortlessly navigate through life and achieve your desired outcomes. An idea is essentially a manifestation of what the subconscious mind believes to be true. Your beliefs form your truth, and once they are embraced, they become the truth of your purpose. Anything you've considered the truth in your purpose is just an idea in disguise. Ultimately, there is no truth except for what the subconscious mind acknowledges as true. Henry Ford once said, “If you think you can, or you think you can't—you're right. But

I say:

- If you think you can't, you're right—and that's your truth.
- If you think that you can, you're right—and that's your truth.
- If you think they won't listen to you because you're a woman, you're right—and that's your truth.
- If you think that all men are sexist, you're right—and that's your truth.
- If you think that all women are too emotional, you're right—and that's your truth.
- If you think there are no opportunities out here, you're right—and that's your truth.
- If you think no one wants to see you employed, you're right—and that's your truth.
- If you think the world does not want to see you win, you're right—and that's your truth.
- If you think you will not get that contract, you're right—and that's your truth.

Your truth does not change the world. It only changes the way the world interacts with you. Change your reality by changing the ideas that shape it. What you believe as "truth" impacts your thoughts. Your thoughts impact your emotions, your emotions impact your behaviors, your behaviors impact your actions, and your actions impact your results. When you alter the underlying ideas, you'll see a difference in your outcomes. By declaring *this idea* or *that idea* as your truth, you set the rules of the game. Too often, organizational leaders, entrepreneurs, and business owners are held hostage by ideas from 25 years ago. I've seen 40-year-old business owners living up to the best thinking of their teenage selves, and they've never challenged these truths as possibly being incomplete or outdated or updated their mental software. Remember this: if the truths you're telling yourselves or the rules you've set for yourself no longer serve your best interests. You are empowered to change them by adapting a new truth.

Your reality, including how you interact with your environment, is shaped by the ideas you accept as truth, so be intentional about what you believe. If your current reality isn't what you desire, it's time to change it. Embrace new beliefs and watch as your reality shifts to align with new, more accurate information. Don't be like the 98% of functional leaders who say:

- *"This couldn't be done."*
- *"They'd never take me seriously."*
- *"All men are trash."*
- *"There are no opportunities."*
- *"No one wants to see me win."*
- *"I won't get the contract."*

They may not realize that the reason the desired outcome did or did not occur is because their subconscious mind has been conditioned to believe that their idea is true, influencing the universe to bring about the desired results. Once something happens that they believe *proves* they're right, they feel justified. This is a dangerous mental space to be in because, too often, people make themselves right by their own standards. They make these truths *their experience*. This is how your mind can deceive you; it gives you a false sense of confidence in your beliefs. When your thoughts become reality, it reinforces your belief that what you're thinking is valid and correct. It becomes hard to present new information and the possibility that they could be wrong. *So, just because you think you're right, you are right!*

Your Rules Should Lead to One Place

It's up to you to set the rules for your game. You are empowered to modify the rules to ensure they lead to your desired outcome or reality. You should reconsider and revise your rules if they're not working. Make sure that the ideas you're putting in place are not just creating a small structure but rather building towards the magnificent outcome or reality you want.

You Determine the Game Rules

The concept of truth is simply a perception. Your subconscious chooses to believe the only truth that exists. We, as humans, can accept and change what we consider truth. This is why it's important to pay attention to the game and keep score. If what you have accepted as truth is not working for you, change the source of your beliefs. Changing your beliefs can alter the outcome of your reality. Because it is the belief that drives all actions. The truth you accept must be carefully chosen. All your beliefs should align toward a common goal. You can base your beliefs on trial and error and assumptions, or you can base your beliefs on universal truths that are tested and proven to deliver consistent results. The choice is yours. This is your game. You set the rules for your game, but the rules you choose to play by are only as beneficial as the results they produce.

Mindset Mechanics Rule #3: You cannot have what you need.

I first learned this growing up in a single-parent household. I noticed the miracles and sacrifices my mother made for me and my siblings. While providing for her children, I would always pray that our needs would be supplied and her struggle would be removed. Many of us have found ourselves in situations during childhood, adulthood, or in business where the need exceeded our capacity to obtain. The truth is, in most cases, no matter how much we willed or hoped for our needs to be fulfilled, we either never received what we wanted or it came with great.

Why do those who need the most have the least, and those who need the least have the most? The person who reads 100 books never reads enough, but the person who finds 10 really good books and continues to reread them and gains more knowledge.

Needing signifies importance.

If you own a business and you're stuck at $50K per month, you're likely feeling frustrated and eager to reach $100K per month. The key is to understand anything important. You create the space for it in your life. What matters to you gets time and priority.

The level of importance determines the pace at which you achieve your desired results. Have you noticed that when you need something, it seems to get further away from you? On the contrary, those who don't need anything seem to always have more than they want. Being in need separates you from being in supply. Need to rearrange your priorities and refocus on what's really important—being the person who already has—before you receive. To have what you need, shift your mindset. To remove the gap between you and what you desire, immerse yourself in what matters most—the feeling of having your desire now. If you believe you already have something, you'll speak and behave a certain way. It's what I call the *Be, Do, Have* philosophy. The state of *being* someone who already has is the only thing that matters. Believe that you already have everything you need before you receive it, and then behave accordingly. Your mind doesn't know the difference, so you resonate as the person who has, and when you look up, what you need presents itself to you. You start acting like someone who has what they need, and it appears. This brings a new meaning to Matthew 6:32-33 NIV, which says:

But your heavenly Father already knows all your needs. Seek the Kingdom of God above all else and live righteously, and he will give you everything you need.

Once you stop *needing*, you'll start *receiving* what you truly want. It's all about becoming the person who already has the frequency to possess everything they need.

Mindset Mechanics Rule #4: The strongest frequency wins.

The power of frequency is undeniable. In ancient times, being the strongest physically was crucial for survival. However, as humans have evolved, so has our awareness. The rules of the game have shifted, and physical strength is no longer the be-all and end-all. We live in a world where frequency is the ultimate decider.

The individual with the strongest frequency wins. Being the person with the highest frequency means holding firm to the belief that your desired reality already exists. Jeff Bezos is a perfect example of this—despite being told that Amazon wouldn't make any money when he first started, he maintained his conviction and is now considered one of the richest men in America.

The world will try to make you lose conviction and doubt your desired reality, but you must stay strong. If you have a vision for your life, you need to maintain that vision and not be swayed by the physical world. This is how you can maintain the highest frequency and achieve what you desire. The top individuals in any field are those who maintain the strongest frequency and conviction. It's not about intellectual or spiritual consciousness but the strength of your heart. Your belief in your desired reality is key, even if your current circumstances say otherwise. If you can maintain this conviction, you will be able to emit the strongest frequency and make your desired reality a reality.

Mindset Mechanics Rule #5: Frequency has polarity.

Understanding the frequency of our thoughts and emotions is critical. In From Passion to Peace, clergyman James Allen helps us achieve our desired outcomes by operating at the highest frequency. A successful mindset vibrates at a high frequency, thus attracting the reality we seek. We can only manifest our desired reality if we are already in a state of mind that aligns with it. It's about tuning our vibration to match the frequency of what we want to attract. This frequency is what draws our desired reality towards us.

How to Change Your Situation with Frequency?

1. **Transcendence** involves refraining from reacting and passing judgment. Your goal should be to respond instead of reacting.
2. **Transmute** is about polarity by moving from the undesirable reality to neutrality. Then, change the frequency to a higher vibration.
3. **Translating** the desired state requires staying committed to the actions that contribute to obtaining your desired outcomes.

Transcendence is not a rare occurrence; rather, it is an integral aspect of personal evolution, and for the 2% of vital mechanical operators, this separates them from the 98% of organizational leaders. Evolution takes the 2% of vital mechanical operators on a polarity-laden journey through self-development—from negative to neutral and then positive desired outcomes. Positive personal outcomes spill over into achieving positive business outcomes. These few leaders vibrate with a higher frequency because they've learned how to convert low vibrational energy into high vibrational energy. They transform negative challenges into positive results. Although only a handful of organizational leaders have achieved this state so far, over time, my mission is to help as many organizational leaders who are ready to eventually enter into it.

Those who ascend to this higher plane find themselves free from judgment, sorrow, and disturbance. Their thoughts are pure, their actions are virtuous, and their lives unfold with a serene rhythm. They have triumphed over their own desires and embraced the essence of truth. They have vanquished undesirable outcomes and have come to understand the true meaning of "goodness." From this point onward, physics will guide us. If you pay attention to universal laws, the embodiment of truth will reveal itself. The ultimate aspiration is to live a life marked by transcendence.

In the journey to transformation, there lies a pivotal phase known as transformation. This phase exists between the torment of undesirable outcomes and the tranquil peace of desired outcomes, embodying a neutral state of change. It serves as the crucial link between the invisible and the physical realms of human experience. The adept mindset mechanic facilitates this connection. As you undergo transmutation, there is a moment when your undesirable outcomes begin to fade away, and true "goodness" starts to flourish. This shift is a profound awakening. Within your mind, you receive a fresh vision, elevated consciousness, and the emergence of a new identity. Upon reaching this milestone, the leader transitions from being a mere human into a formidable authority. The leader has been "born again," embarking on a fresh series of experiences that expand their horizons to apply the Midas touch to their organization, teams, and results. With this transformation, newfound powers unfurl, revealing a vast and intricate awareness of invisible forces. It is within this exquisite state that the zenith of transcendence is ultimately attained. This, then, is the essence of what James Allen called the Transcendent Life.

Once translation is truly embraced, life unfolds in a way that feels right. Your state of being mirrors the natural condition of your future desired state. The external attacks, relentless pressures, and injustices that once inflicted pain on you seem to amplify, but instead, it's only as joy. Rather than diminishing your undesired state, challenges motivate you, and they are turned into even greater richness to feed your desired outcomes.

The most common error is attempting to change a situation without considering the order and then applying principles to change your inner universe. You must elevate yourself above judgment, transform your outlook, and then communicate your frequency and desires so others can see your best work and high-performance results.

Mindset Mechanics Rule #6: Universal laws won't respond to your definition of "goodness."

In the world of law and physics, being a "good" person doesn't matter. The universe only responds to frequency. As leaders, we all want to be seen as kind and decent to others, but the truth is that the universe doesn't take humanitarian goodness into account. This type of goodness only has value among other humans. The universe only responds to obedience to universal laws and behaviors that match a specific cadence, resonance, and frequency. While it's important to be a *decent* person on a humane level, when it comes to achieving your goals, humanistic goodness has no value. To get your desired results, you need to appeal to universal laws that embody a completely different definition of what *goodness* truly is. At the highest level of existence, concepts of morality don't hold sway. This is exemplified by the fact that nature's laws do not change to save innocent children. Bobby Hemmet urges us to think like gods, understanding that it's our frequency, not our humanity, that brings about results. As leaders, our aim should be to align our humanity and creativity to achieve our desired outcomes while also fulfilling the needs of those around us.

Mindset Mechanics Rule #7: What you resist persists.

Resisting grace only allows problems to persist; while embracing it leads to a life filled with blessings. Grace is a valuable gift but often goes unnoticed when we push back against it. Rejecting grace means missing out on the potential it holds and settling for less. It's important to understand that grace is a gift that we can either accept or settle for. Settling for less leaves us feeling unfulfilled and stuck in a cycle of disappointment. You are empowered to choose the gift of grace or settle for less, but why not strive for the best? My mission is to provide Mindset Mechanics as a valuable tool to help transform startup and growth-based businesses and personal brands into more profitable and enjoyable ventures.

Mindset Mechanics Rule #8: Your problems are not outside of you; they are inside.

When you're experiencing internal conflict while striving to achieve your objectives, it's not external factors holding you back; it's something within you. There are both contributing and determining factors. Contributing factors are external, while determining variables are internal. For example, in 2024, many people claim that racism is preventing them from succeeding. Racism is an external factor; it might slow you down, but it cannot stop you from reaching your goals. Similarly, some women feel hindered by being in a male-dominated environment. While this may pose challenges and slow progress, it does not prevent success. These external realities are contributing factors, not determining ones.

Mindset Mechanics Rule #9: A strong mind maintains positive thinking.

How many times in the last day have you found yourself dwelling on thoughts of scarcity? How many times have you worried about not having enough, running out, or not being accepted? Do these thoughts reflect an attitude of abundance, or are they a sign of neglect? It's possible that, despite all your hard work, your default mode causes you to rely too heavily on physical labor. This prevents you from exerting the mental effort required to become more diligent in your thinking. A leader's goal is to be mentally diligent so that you don't have to rely solely on physical effort.

If you neglect to exercise your mind, you'll end up exhausting your body and ruining your health. If you're not used to thinking deeply to find solutions, when the time comes to make a critical decision, your thoughts will lean towards scarcity, filling your mind with doubt. But this conclusion doesn't come from reality; it comes from your mindset. My whole perspective changed when I realized that we don't see things as they appear to us. Like me, you'll start to notice when you're making a situation seem more difficult than it is. Are you letting yourself off the hook by thinking, "It's not my time yet" or "This can never happen for me"? What words are you using to excuse yourself from finding a solution? Today, I challenge you to focus only on diligent choices because they lead to abundant outcomes. It might take time, but your mindset will grow and empower you.

Mindset Mechanics Rule #10: Pressure creates diamonds.

Oftentimes, we are quick to act to relieve our physical and mental pressures. We decide that the task at hand is too daunting and costly, so instead of making the necessary changes, we find it easier to avoid change entirely and simply adapt to the status quo without having to put in any effort to solve problems. Change is less expensive than staying the same in the long run, and the toll it takes on us can be incredibly high. Just think about how often we purchase things we "can't afford" simply because we want them. We know we need to make changes, but we are afraid of failing, so we choose to avoid the work that needs to be done. What if people bought what they truly needed rather than what they wanted, both consciously and unconsciously? To overcome our mental pressures, we must be willing to do what needs to be done and endure the small costs of change to come out on the other side.

Mindset Mechanics Rule #11: Language frames your worldview.

The language we use and the thoughts we have directly impact our outcomes. Our words and thoughts can either empower us or hold us back from taking action. If you're not achieving your desired results in life or business, it's crucial to analyze the language, perceptions, thoughts, and overall culture of those around you. Whether you realize it or not, your beliefs shape your life and business.

Mindset Mechanics Rule #12: Quality attracts quality.

It's important to be deliberate about what you welcome into your life. Consider your clothing, car, and food quality. The way you handle one part of your life influences all the others. Take note of both the standard and how much you have. Are you a quality collector or a quantity purchaser? When I started emphasizing quality in my transactions, I discovered that I required fewer things, and what I bought lasted longer and brought more satisfaction. Life recognizes you based on the quality, not quantity, of the value you produce. A high-quality result is called value. Creating value only comes from high-quality energy sources. If you want value in your life and business, you must obtain new energy from high-quality sources.

Mindset Mechanics Rule #13: Wealth is spiritual.

I have come to understand that true wealth is not simply measured in money but rather rooted in spirituality. Putting in hard work for profits is about materialism, whereas growing true wealth is about expanding beyond physical labor to maximize your income potential. For leaders to effectively manage people, it requires strong social skills and patience, while honing communication and finding your voice has the potential to yield significant financial gains. The ability to imagine and think creatively represents the pinnacle of true wealth. Wealth comes from creating something valuable for others rather than oneself. I have decided to use my mental energy and voice to develop solutions for others, leaving behind a legacy and empowering others to do the same. Formulating solutions is a personal obsession of mine, and helping others is my driving passion. My life is dedicated to Mindset Mechanics so that you can focus on building the life and business of your dreams!

Mindset Mechanics Rule #14: Begin with the end in mind.

Everything must have a clear purpose and intention. For example, we create businesses for profit. We document and develop business systems exclusively for delegation. If you are a solo entrepreneur, why are you building a business or documenting your processes? Hopefully, you are creating business systems and documenting your operating procedures to either grow your business and hire employees or increase the value of your business for future sales and acquisitions so the buyer can delegate the business systems to their employees. Business systems' sole purpose is delegation.

Mindset Mechanics Rule #15: Be hyper-intentional about your thoughts.

Your mind is your most powerful asset—it can either propel you forward or hold you back. As a leader, developing the right mental framework is crucial for achieving your vision. But this isn't just about thinking positively; it's about being *hyper-intentional* with your thoughts, shaping them in a way that aligns with your goals. In *Mindset Mechanics,* we dive deep into the process of refining your thinking to elevate your performance and drive meaningful change.

To expand your mental capacity and think at a higher level, you must be deliberate about your thoughts. The mind is incredibly powerful, but it can also prematurely assess the difficulty of a situation without evidence. As a leader, you need to be aware of the mind's many functions and avoid making snap decisions. Take the time to categorize your thoughts, analyze their impact, and determine how they align with your desired outcomes. Here is a brief understanding of how your mind works so you can be more intentional about identifying which mental capacity you operate in most.

Imagination is a powerful tool that lets you envision outcomes and ideas before they materialize. It's like a superpower for the mind, allowing you to see possibilities long before they become reality. Leaders and creators of all kinds depend on imagination to bring their ideas to life—it's the foundation of all great achievements.

1. *Cognition* is the foundation of awareness. From a young age, we begin to recognize familiar faces, body parts, and surroundings, gradually distinguishing between left and right hands.

2. *Correlation* is the ability to identify differences, like understanding that a hand is not the same as a foot.

3. *Comprehension* goes beyond awareness to grasp the deeper workings of something. For instance, when I played the alto saxophone in high school, I truly understood the relationship between finger placement, musical notes, and scales. It wasn't just memorization; it was genuine comprehension. Many people hold knowledge without truly understanding it.

4. *Memorization* allows us to retain information by heart, saving time by making it easily retrievable without constant review.

5. *Prioritization* directs our focus toward key tasks. Anticipating positive outcomes can motivate us, while anxiety depletes our energy by focusing on potential negative scenarios.

6. *Categorization* involves organizing ideas into specific classifications and understanding their appropriate time, place, and purpose.

7. *Rationalization* often uses logical excuses to avoid change, convincing ourselves that certain outcomes are impossible and accepting that as

truth.

8. *Distortion* occurs when we interpret information to fit our existing beliefs rather than the reality of the situation.

9. *Generalization* leads to oversimplification, often using labels to discredit opposing viewpoints. This tactic, known as ad hominem, aims to undermine those with differing opinions, weakening their arguments by reducing them to a single negative label.

For a long time, I was a functional thinker, convinced that only my actions could change my results. I wanted to think smarter and not work harder, but I didn't know how. I kept reverting to my old programming of pushing hard because it was all I knew. While it brought results, it also left me depleted. When life-changing advice conflicted with my values, I dismissed it—rationalizing, generalizing, or distorting it—blocking any chance for change.

I didn't realize how powerful my thinking was. I neglected my well-being, lacked mental discipline, and couldn't discern the value of sound guidance. I had no reference point for making better decisions, which left me financially strained, misinformed, and fearful for too long.

Success offers valuable lessons, and one of the most important is not to assume the difficulty of something before trying it. Once I became more intentional about my desires and recognized the mind's various functions, I evolved into a leader positioned for success.

Mindset Mechanics Rule #16: Your voice is powerful.

Our voices are powerful. It shapes our reality through the declarations we make. Our words can bring about change, impact our frequency, and direct our intentions. That's why it's crucial to be mindful of our language, so you can resonate and be seen and heard as the solution the world needs most. The right mindset is the first step toward achieving success, and the rest of this book focuses on practical strategies for success in your business or organization.

Now that you've read the 16 rules of *Mindset Mechanics*, let me ask you: what truly drives people to change their thinking? In *Influencer: The Power to Change Anything*, the authors explain that, for change to happen, people *have to want it*. I agree. Real transformation starts with a burning desire ignited from within. While some are driven by fear or peer pressure, the top 2 percent of leaders are motivated by love.

Remember Mrs. Marva Collins, founder of the Westside Preparatory School in Chicago? She was featured on *60 Minutes* and was even offered the role of Secretary of Education by President Reagan. Mrs. Collins stayed committed to her students, believing in their potential long before they did. Every morning, she greeted them with a mantra: *"Welcome to success. Say goodbye to failure because you are not going to fail here. I will not let you fail. You are here to win. You were born to win."*

Her principles were simple but powerful: teachers were the "little U's" serving the "big I's," the students. Teachers didn't sit at desks; they walked the classroom all day, helping students discover their uniqueness and potential. By fostering an environment of mutual respect and individuality, Mrs. Collins mastered her language and was able to create true inclusion and diversity. She found her voice and used it to emphasize the basics of reading—phonetics, patterns, and principles—to unlock each student's hidden genius.

Mrs. Collins' passion for excellence drove everything she did. She famously said, *"Society predicts, but I determine."* Her life's work exemplifies that love is the most powerful motivator—it sustains drive, focus, and perseverance. When you root your desires in love, you spark meaningful change in your organization.

Find a cause worth pursuing, and let love guide you, even when the road gets tough. Just as Mrs. Collins did, allow your passion to align with everything else.

---

21. Kahneman, Daniel. *Thinking, Fast and Slow*. Farrar, Straus and Giroux, 2011.
22. *Success! The Marva Collins Approach (1981)* Documentary. Youtube video, https://www.youtube.com/watch?v=yXIDVjDlXpc

23. *60 Minutes*, "Marva Collins 1995 Part I," Youtube video, https://www.youtube.com/watch?v=h8b1Behi9FM
24. Allen, James (1901). *From Poverty to Power*, p. 16.
25. Nobre, F.S. and Steiner, S.J., 2003, *Perspectives on Organizational Systems: Towards a Unified Theory*, Scientific A Model of an Organizational System. Figure on ResearchGate. Available from: https://www.researchgate.net/figure/A-Model-of-an-Organisational-System_fig1_277302639 [accessed 17 Sep 2024]
26. Harrington, Kevin, and Daniel Priestley. *Key Person of Influence: The Five-Step Method to Become One of the Most Highly Valued and Highly Paid People in Your Industry*. Rethink Press, Limited, 2015.

CHAPTER 11. The Business Vehicle

In *Rework*, Jason Fried and David Heinmeier Hansson define success as "the experience that actually counts." They emphasize that failure is often part of the entrepreneurial process, with advice to "fail early and fail often." Many say nine out of 10 new businesses fail, which might seem discouraging. But don't let these statistics deter you. I believe past failures don't define your success; they're actually indicators of future success. What truly determines success or failure is your *Business Vehicle*.

Struggling to market products, build a team, or price services correctly doesn't mean you're a failure—it just shows a greater need for learning and understanding why businesses thrive or falter. Instead of dwelling on the negativity surrounding failure, focus on your ability to learn and try again.

Next, we'll dive into the specifications of your *Business Vehicle*, which represents your organization (see **Figure 14**).

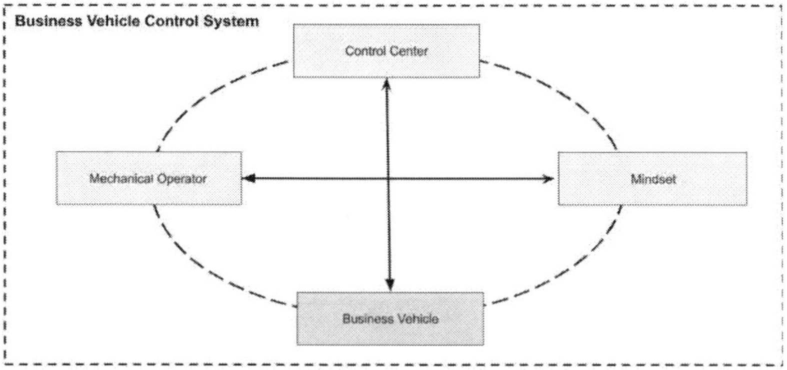

Figure 14. Business Vehicle Control System: Business Vehicle

There's a common belief that you must learn from your mistakes. But what's the real value of focusing only on mistakes? Why not learn from success instead? Building a successful *Business Vehicle* isn't just about avoiding pitfalls—it's about understanding how the right design propels you forward.

At the heart of every business venture lies a fundamental question: What makes something fail or succeed? Imagine if there were a universal answer. Would you implement it? Absolutely. It wouldn't just highlight what you're doing wrong; it would validate what you're doing right and guide your future decisions.

Learning from success offers tangible proof of what works and gives you the tools to replicate and enhance those strategies. It refines your methods, builds your skills, and sets the stage for even greater achievements.

Let's explore what makes a successful *Business Vehicle.*

Rework Your Business

Sometimes, you need to start over and rework your business. Perhaps you chose the wrong industry, clients, team members, strategies, or outdated processes. But don't be discouraged—failure doesn't disqualify you from success. In fact, a Harvard Business School study found that entrepreneurs who have previously succeeded are more likely to achieve success again, with a 34% success rate for their future ventures. Meanwhile, those whose companies failed initially have a similar success rate as first-timers—about 23%.

Failure is not a setback but a step forward, aligning with the way nature operates. The universal laws of physics don't dwell on past failures; they constantly build upon successes, and so should you.

There's one factor that ultimately determines success or failure: *performance.* It applies to individuals, businesses, communities, and even NFL teams. In building your *Business Vehicle,* focus on performance. This key element is part of a simple, powerful formula for success. And the best part? You don't need an engineering or physics degree to understand it—once you grasp it, you'll see its relevance in every aspect of life.

The Machine Matters

The *Business Vehicle* you choose matters. You have several options for traveling the Road to Success: on foot, by bike, car, airplane, jet, boat, or rocket ship. The key is selecting the right vehicle for the journey because *time* is your main barrier. Traveling by foot may take a lifetime, so be prepared to pass the baton to the next generation. If you opt for a rocket ship, you'll need skilled rocket scientists and engineers—and be ready to compensate them well, as they are rare.

For most of us, using a car as our *Business Vehicle* is ideal. It's not too slow, has just the right acceleration, attracts the right customers and team members, and will get us to our destination. Whatever vehicle you choose, make sure it aligns with your goals and needs, as the *Universal Systems Model* of *Mindset Mechanics* still applies.

When selecting your *Business Vehicle*, you also have choices in features: engine size, transmission type, seating capacity, body style (coupe, sedan, or SUV), brand (e.g., Mercedes), and fuel type (electric, regular, or premium). Choose features that offer style, comfort, and performance—no one wants to ride a struggle bus.

Your *Business Vehicle* should align with your cultural hub, fulfill your vital needs (happiness and energy), and set high-performance standards. It must have no energy leaks, low chaos, adequate seating, and use high-quality energy sources. Every component must be properly fitted to build momentum and accelerate without falling apart. In essence, your *Business Vehicle* must align with universal principles.

The Universe is Systems-Minded

As an inquisitive child eager to understand how things worked, I believed I needed to break them into individual components. But over time, I learned that understanding the parts isn't enough; you need to see how they function as a *complete system*.

A system is a set of interconnected components working toward a common goal. Successful organizations operate like systems, following the laws of thermodynamics. They understand that quality drives the system and that conserving energy and building mass is essential. According to the second law, internal energy is never 100% efficient. Some energy fuels the system, some is lost to chaos, and all systems eventually need replenishing. Regardless of how well-designed they are, systems break down over time due to aging—a concept known as the *arrow of time*. As Stephen Hawking once put it:

> The increase of disorder or entropy with time is one example of what is called an arrow of time, something that gives a direction to time and distinguishes the past from the future. There are at least three different directions of time. First there is the thermodynamic arrow of time—the direction in which disorder or entropy increases. Second, there is the psychological arrow of time. This is the direction in which we feel time passes—the direction of time in which we remember the past but not the future. Third, there is the cosmological arrow of time. This is the direction of time in which the Universe is expanding rather than contracting.

Even humans experience disorder as we age; things start to change or break down. Successful organizations understand that *energy management* determines results, making it crucial to minimize wasted energy to maintain efficiency and optimize performance.

The goal of any successful system is to balance *input* with *output*. In thermodynamics, chaos is referred to as *disorder*. While some system disorder is inevitable, the aim is to manage them effectively. Mismanagement of disorder leads to system failure, so systems need consistent maintenance or repair. Low-quality energy and poor management result in breakdowns. To truly enhance a system's performance, you first need to understand it.

Every element, big or small, belongs to a *complex adaptive system* (CAS). Whether it's the galaxy, the human body, or a transportation network, each operates as part of a larger, interconnected system. Systems consist of interdependent subsystems, all communicating and working together. It's vital to recognize and appreciate the complexity, independence, and interdependence of these systems.

To improve performance, remember that *quality is a function of your system's input*. Simply put, your outputs can never exceed the quality of your inputs. Also, you cannot control the system's output, you can only control system inputs. The ultimate aim of any system is to achieve efficient outputs, which depend on effectively managing inputs according to the *Laws of Thermodynamics*.

All Systems Follow the Laws of Thermodynamics

Two apply when seeking to optimize a system. In physics, these laws are called:

1. **First Law of Thermodynamics**
2. **Second Law of Thermodynamics**

The *First Law of Thermodynamics*, also known as the *Law of Energy Conservation*, is often called the *BIG Energy Law*. Simply put, this law states that energy cannot be created or destroyed—it can only be transformed from one form to another. The energy and mass that enter a system must equal the amount that exits (see **Figure 15**).

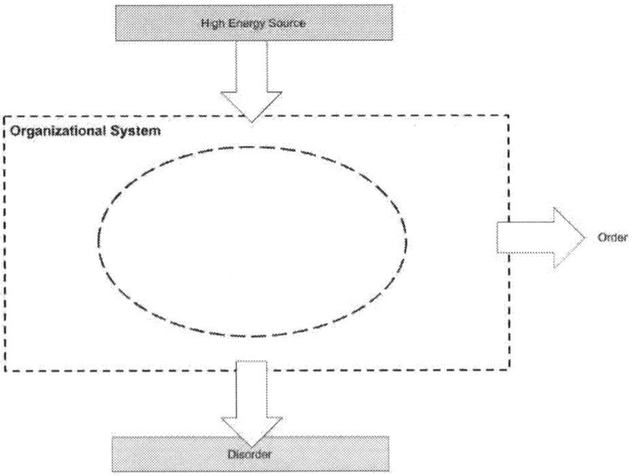

Figure 15. First and Second Laws of Thermodynamics

The energy needed to power your business already exists and is always seeking equilibrium. It's readily available in the environment—waiting for someone to harness, utilize, and manage it effectively to yield positive results. The fundamental principle of the *BIG Energy Law* is that energy naturally flows from higher sources to lower ones, not the other way around.

The *Second Law of Thermodynamics*, the *Law of Order* in Energy Processing, explains that energy is first used to meet internal needs before contributing to performance. An organization's performance depends on energy quality and system efficiency. If you're burned out before you start, your performance suffers. If an organization consumes too much internal energy, it won't have enough left for high performance. High-quality energy sources boost performance, while excess internal energy leads to undesirable outputs like entropy, waste, or chaos. In systems theory, managing these unwanted outputs is essential because they are always present and expanding in the universe.

The Laws of Thermodynamics are why energy management is so important. In systems, as well as in business, the goal is to control energy loss and overconsumption by preventing internal system failures and good management practices so you can increase system success. This is achieved by selecting the best systems training, designing the best system, and sourcing the best material and energy possible for your organization.

The specifications for your business vehicle must follow the first and second laws of thermodynamics. Combined, thermodynamics states:

1. Energy is finite, so new energy (ideas, results, and solutions) is needed for a system, but it can only be accepted from high-energy sources within the environment. A problem cannot be solved using the same energy source that created it. You must seek new energy from an eternal, higher source to increase the probability of sustainability, longevity, and performance.
2. It also states that system behaviors are predictable and orderly, first meeting the organization's internal needs before creating desired results.
3. Finally, it states that a system requires internal and external management and maintenance. If a system fails to manage its

energy so that it cannot cover its internal needs, it will experience mechanical failure. If a system is not properly maintained, it could experience permanent failure.

Next, we will review the secret formula for creating and maintaining a high-performance culture. This formula calculates how well a system manages its internal energy. Failure to manage internal energy rapidly increases the ever-present chaos phenomenon called disorder or entropy.

---

27. Berlin, Leslie (March 21, 2009). Try, Try, Try, Again, or Maybe Not. *New York Times.* https://www.nytimes.com/2009/03/22/business/22proto.html
28. Eddington, Sir Arthur. *The Nature of the Physical World: The Gifford Lectures 1927.* Edited by Klaus-Dieter Sedlacek, BoD—Books on Demand, 2021.

CHAPTER 12. The Mechanical Performance Formula

Have you ever wished for a formula that could simplify your path to success? Imagine a clear, mechanical solution that ensures you're (1) in the right environment, making the right decisions, and delivering the right value, and (2) achieving your goal to align growing opportunities with differentiating capabilities that drives your efforts forward, maximizes efficiency, and minimizes frustration. As business leaders, we often seek a direct path—one that not only saves time and resources but also produces remarkable results with less effort. If such a formula existed, wouldn't you want to know how to apply it?

That's precisely what the *Mechanical Performance Formula (MPF)* offers. Earlier, we established that your mind operates like a machine, and focusing on the right things can lead to significant breakthroughs in life and business. This formula directs your efforts, automates processes, and is a hidden superpower for achieving your desired outcomes.

The *Mechanical Performance Formula* determines the success of any system. When applied effectively, it streamlines decisions, clarifies actions, and sets you up for recession-proof, disaster-proof, and even poverty-proof results. It provides a comprehensive approach to understanding business systems—how they function, how to optimize them, and how to predict outcomes. With this formula, you move beyond trial and error, gaining insight into the mechanics of your business.

You don't need to be a scientist to grasp this concept. By learning the fundamental laws of physics as outlined in *Mindset Mechanics*, you'll be equipped to make this formula work for you and your organization. At its core, the formula is straightforward (see **Figure 16** and **Figure 17**):

$$\text{Performance} = \sum \left[\frac{\text{Desired Output}}{\text{Required Input}} \right]$$

Figure 16. The Mechanical Performance Formula

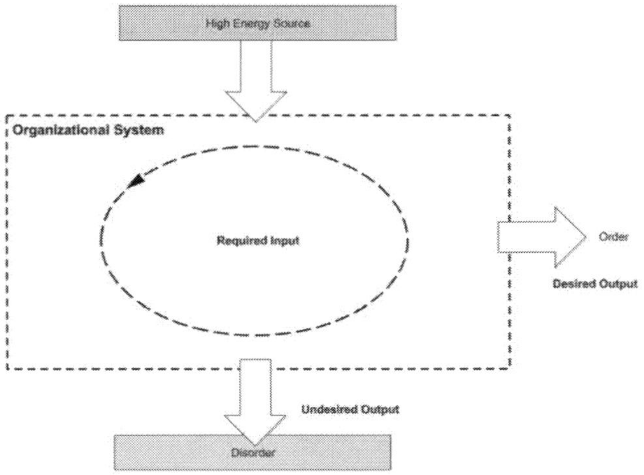

Figure 17. Illustration of How Energy Flows through the Mindset Mechanics Interface.

This simple equation holds the key to achieving efficient results. It helps you distinguish between good and bad decisions, ensuring that every action aligns with your business goals. Let's dive deeper into how the *Mechanical Performance Formula* can transform your approach and accelerate your success.

The *Mechanical Performance Formula* (MPF) is a groundbreaking concept that merges first principles, mechanics, and scientific thinking into a single, powerful equation. It embodies a simplified approach rooted in first-principle thinking, breaking down systems into their fundamental components for a clearer understanding of how they work. The formula incorporates mechanical thinking through its automatic, pre-programmed nature while also embracing scientific thinking to test ideas, process results, and gain insights through observation.

The *MPF* not only calculates an organization's energy capacity to do work but also identifies systemic issues and their origins. It reveals how problems are best solved at the appropriate management level, enabling leaders to allocate energy wisely. As a leader, your energy is one of your most valuable assets, but daily challenges often drain it. When you're constantly putting out fires, how well does your business truly perform?

This raises a vital question: What consumes more energy—thinking or labor? Understanding this distinction is crucial. The *MPF* helps you focus your energy on strategic thinking rather than being overwhelmed by routine tasks, leading to improved performance and sustained growth.

How the Mechanical Performance Formula Works

Now, let's dive deeper into how the *Mechanical Performance Formula* is optimized within a system. The fundamental laws state that:

1. All systems receive available energy from their environment.
2. All systems use available energy first to meet internal needs.
3. All systems trend towards failure over time.

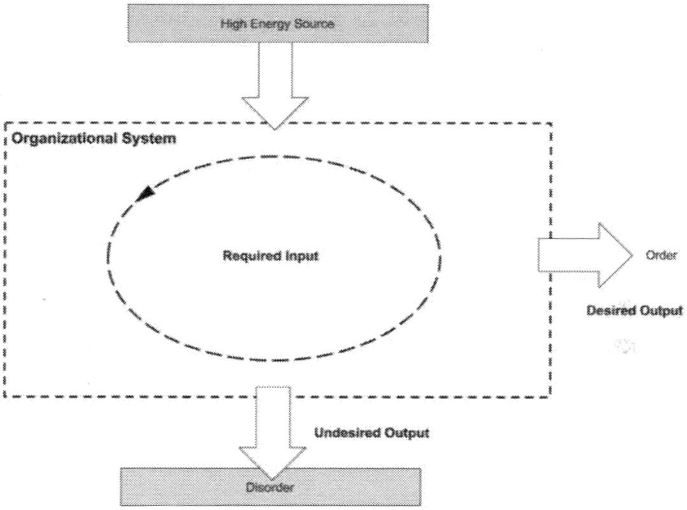

Figure 18. Overall System Performance: Performance = Desired Output Over Required Input. The goal is for Actual Output ≥ Desired Output.

High performance is the ultimate goal for any organization. As shown in Figure 4, disorder is a constant in the universe, and it must be minimized to enhance system performance. As discussed earlier, available energy is first used to meet the internal needs of the system before it can be directed toward achieving desired outcomes.

To better illustrate the *Mechanical Performance Formula* in action, let's explore a few examples of a system's ideal performance, underperformance, and overperformance, along with their impacts.

Ideal Performance occurs when the actual output matches or exceeds the desired output.

Example 1: Imagine 100 points of available energy being supplied to a system that requires 100 points of input. In this scenario, 40 points of energy are lost to disorder or undesired output, while 60 points of energy remain as desired output to perform work. The result? An efficient system where energy is directed toward productive outcomes and performance is optimized.

Understanding these dynamics can help you identify where energy is being wasted and where adjustments are needed to boost performance. By aligning energy inputs with system needs and minimizing disorder, you can consistently achieve high performance.

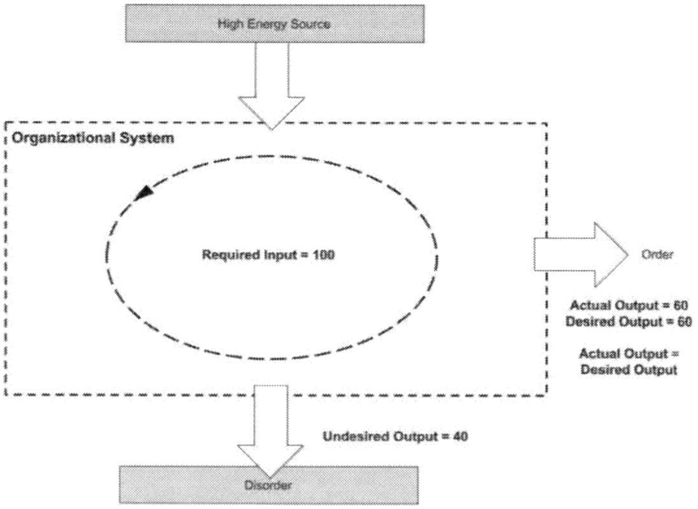

Figure 19. Organizational Ideal Performance (Managed Disorder): Actual Output Equals Desired Output.

Now, the Mechanical Performance Formula is calculated with 60 percent efficiency.

$$\text{Performance} = \sum \left[\frac{60}{100} \right]$$

$$\text{Performance} = 60\% \text{ Efficiency}$$

Figure 20. Ideal performance of a system equation.

Performance measures how effectively a system manages its internal energy. Optimized energy management is key. While no system can ever achieve 100% efficiency, the goal is to maximize performance while keeping undesired output—or disorder—as low as possible. According to the laws of thermodynamics, disorder is inevitable but can be minimized.

So, what exactly is undesired output, or what does thermodynamics refer to as disorder? Disorder represents the inefficiencies within a system—energy that is not effectively converted into productive output. Every system encounters disorder, but the level of undesired output indicates how well the system is maintained and how efficiently it operates.

Disorder is energy that returns to the environment as low-quality, less useful energy. Systems that generate minimal disorder are considered efficient, while those with high disorder are deemed inefficient.

No system is flawless; disorder is always present to some extent. However, excessive disorder can lead to system failure. In vehicle mechanics, disorder appears as unwanted engine vibrations, overheating, premature coolant circulation, improper lubrication, oil leaks, or loss of air pressure. The more disorder you eliminate, the better your system performs and the longer it lasts.

Undesired Output in a System

Undesired output is equivalent to internal failure. There are two main ways this failure occurs: low-quality energy intake or internal system mismanagement. Let's break down both scenarios.

Failure (A): Low-Quality Energy vs. High-Quality Energy

When a system receives low-quality energy from the environment, it struggles to sustain itself. To compensate, resources are reallocated from other subsystems, causing the entire system to become overburdened. The Second Law of Thermodynamics emphasizes that systems should produce work using higher-quality energy sources, as optimized performance directly depends on energy quality.

Failure (B): Internal Mismanagement vs. Proper Management

Mismanagement within an organization leads to internal breakdowns. Gaps, inefficiencies, and leaks require more energy to maintain the system, draining resources and lowering performance. Proper management ensures that resources are used efficiently, preventing internal failures and maintaining stability.

Failure A: Impact of Low-Quality Energy vs. High-Quality Energy

Let's illustrate the effects of low-quality energy.

Example 2: Assume that 100 points of available energy are supplied to a system with a required input of 100 points. Of that, 40 points are lost as undesired output, and 60 points convert to desired output, generating productive work. In this case, the energy quality is only 25%, meaning only 25% of the available energy is useful. The Mechanical Performance Formula demonstrates that low-quality energy impacts overall system performance and efficiency.

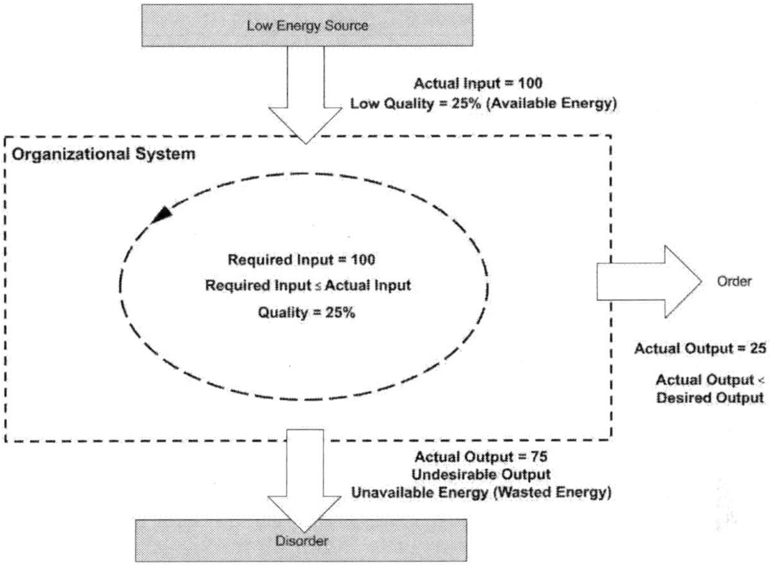

Figure 21. Organizational Performance due to Low Quality (High Disorder): Actual Output < Desired Output.

Now let's calculate the Mechanical Performance Formula:

$$\textbf{Performance} = \sum \left[\frac{25}{100} \right]$$

$$\textbf{Performance} = \textbf{25\% Efficiency}$$

Figure 22. Performance of low-quality energy equation.

These examples demonstrate how well a system manages its internal energy. In the case of low-quality energy, overall performance is only 25%, resulting in underperformance. When available energy lacks quality, it causes systemic issues, leading to lower performance and efficiency. This violates the fundamental laws of thermodynamics, as energy must originate from a higher-quality source to achieve optimal results.

Now, let's consider **high-quality energy**.

High-Quality Available Energy aligns with the laws of thermodynamics, facilitating better system performance.

Example 3: Imagine a system receiving 100 points of available energy with a required input of 100 points. In this case, 40 points are still lost as undesired output, but 60 points are transformed into desired output, generating productive work. Here, the energy quality is 75%, meaning that 75% of the available energy is effective and useful. The Mechanical Performance Formula, in this scenario, reflects a more efficient system that optimizes available resources.

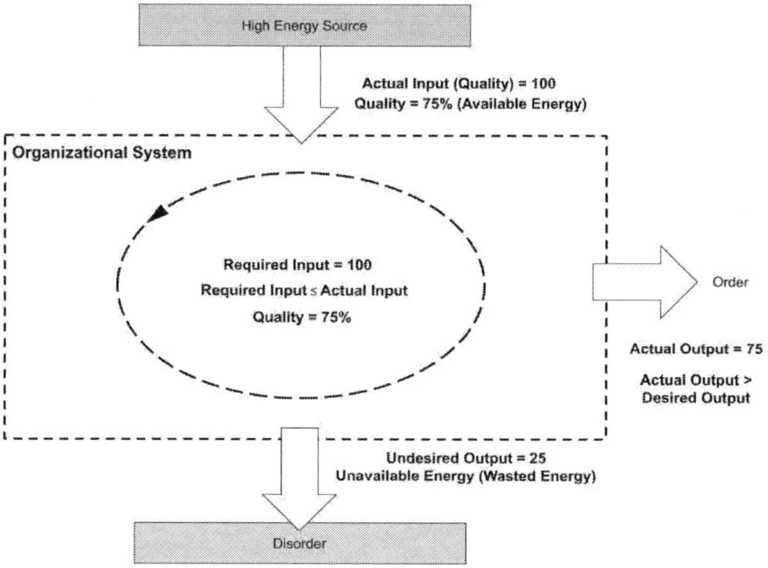

Figure 23. Organizational Performance Due to High-Quality Energy (Low Disorder): Actual Output > Desired Output

Now the Mechanical Performance Formula calculates as:

$$\text{Performance} = \sum \left[\frac{75}{100} \right]$$

$$\text{Performance} = 75\% \text{ Efficiency}$$

Figure 24. Performance of high-quality energy equation.

This example shows how effectively the system manages its internal energy. With high-quality energy, overall performance reaches 75%, indicating that the system is overperforming. But what happens when a system faces internal mismanagement?

Failure B: Impact of Internal System Mismanagement vs. Proper Management

Mismanaged Internal Systems arise when subsystems—such as vision, people, processes, and strategy—are misaligned. In these cases, even high-quality energy cannot prevent failure, as internal leaks disrupt overall performance.

Now, consider a system that receives high-quality energy from the environment but suffers from internal mismanagement or misalignment. This results in excess internal heat, increasing disorder, and undesired outputs.

Example 4: Imagine a system with 100 points of available energy and a required input of 100 points. In this case, all 100 points of energy are lost to disorder, leading to total failure with no productive output. The Mechanical Performance Formula in this scenario reveals the impact of internal breakdowns, where even the best available energy cannot overcome mismanagement.

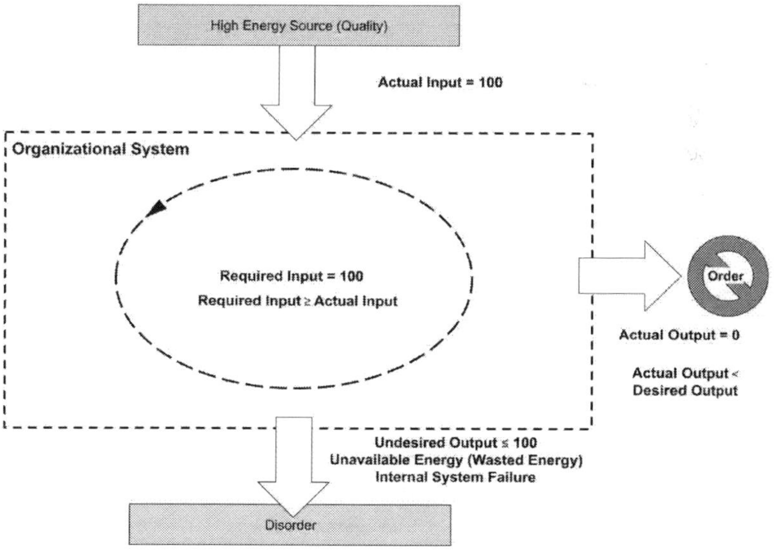

Figure 25. Organizational Underperformance Due to Mismanagement (High Disorder): Actual Output < Desired Output

Now, let's calculate the Mechanical Performance Formula.

$$\text{Performance} = \sum \left[\frac{0}{100} \right]$$

$$\text{Performance} = 0\% \text{ Efficiency}$$

Figure 26. Performance of mismanaged internal system equation.

This example demonstrates how a system manages its internal energy. With internal mismanagement, overall performance drops to 0%. The system essentially fails to function. Now, let's explore what happens when a system is properly managed.

Managed internal systems focus on the interconnections of the system, and subsystems—vision, people, processes, and strategy—are aligned and work cohesively toward a common goal. This results in overperformance, as the system effectively optimizes its internal energy.

Example 5: In this scenario, 100 points of available energy are supplied to a system with a required input of 100 points. Here, 100 points of energy are optimized to produce the desired output, meeting all internal needs with minimal waste. The Mechanical Performance Formula for this managed system demonstrates how aligned components lead to maximum efficiency and effectiveness.

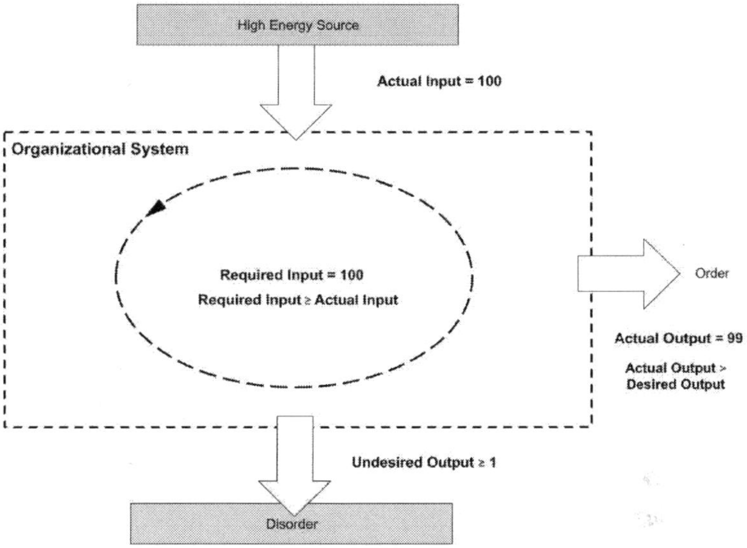

Figure 27. Organizational High Performance Due to Management (Low Disorder): Actual Output > Desired Output

Now, let's calculate the Mechanical Performance Formula.

$$\text{Performance} = \Sigma\left[\frac{99}{100}\right]$$

$$\text{Performance} = 99\%\ \text{Efficiency}$$

Figure 28. Performance of managed internal system.

As the examples demonstrate, system performance depends on how well available energy is converted into ordered work. Adding more resources alone won't boost performance; what truly matters is how effectively you manage the energy you already have.

The Mechanical Performance Formula emphasizes that organizational leaders must grasp how efficiently their systems operate. With this knowledge, they can optimize the system to manage energy more effectively, minimizing chaos and waste. Otherwise, resources will be squandered, leading to eventual failure.

Order reflects how much of the energy supplied from the

environment is converted into useful work. In the universe, available energy is finite and ranked by quality—just like time, space, resources, and money. Not all businesses, systems, or energy sources are created equal, and leaders must learn to evaluate energy quality.

The system's ability to absorb and utilize energy efficiently depends on the compatibility between the system and fuel quality. Just as you wouldn't use premium fuel in a Ford Fusion or regular fuel in a Mercedes-Benz C-300, in business, you must align service quality with cost and customer experience. For instance, purchasing new equipment but charging low prices or using outdated equipment but overcharging clients disrupts the balance.

To maximize efficiency and minimize waste, feed the system with the right fuel quality and quantity. When properly aligned, the result is optimized energy management, reduced waste, and better performance.

Examples of Undesired Output

When systems are well-organized, internal energy is directed first toward meeting organizational needs. However, systems don't always perform as intended. To better understand **undesired output**, let's explore a few real-world examples:

1. **Emotional Leaks**: Feelings of sadness, anger, or betrayal create disorder, leaving you mentally distracted and drained. Your emotional state is subject to the laws of thermodynamics—leaks of energy slow down your productivity.
2. **Healing in Hospitals**: After a heart attack, patients typically rest for extended periods to focus their energy on healing. Hospital staff often limit visits to ensure the patient conserves energy, demonstrating how energy is allocated for internal recovery.
3. **Team Dynamics**: In the NFL, you'll find two types of influential players. On one hand, there's Tom Brady, a world-class quarterback who elevates team morale and performance. On the other hand, there are players like Antonio Brown whose behavior

creates disorder in locker rooms. Despite his athletic talent, his inability to avoid conflicts led to team dysfunction and eventual dismissal. A wise coach knows to avoid players who bring more chaos than cohesion.

4. **Family Conflict**: Financial struggles cause disorder in a family, leading to less energy available for paying bills, meeting needs, or resolving conflicts. Stress increases, and the family's quality time and effectiveness diminish.
5. **Corporate Dysfunction**: When a company's executives are at odds, it leads to high turnover, silos, stagnation, mistrust, and poor communication—affecting sales, strategy, and teamwork. This type of dysfunction results in energy leaks that weaken overall performance.
6. **Government Efficiency**: A disorganized government exhibits disorder in the form of micromanagement, partisan politics, shutdowns, finger-pointing, and a lack of long-term planning. These inefficiencies lead to poor management of programs and services.
7. **Global Issues**: Inflation is a glaring example of global disorder. For 28 consecutive months, according to IPSOS, global inflation has been accelerating. Post-pandemic demand has outpaced supply, driving up costs. As the cost of living rises, poverty levels increase, creating further disorder worldwide. Even with efforts by the World Bank to redefine poverty lines, the rising cost of energy continues to drive global economic instability.

Understanding these examples of undesired output can help leaders identify and manage internal energy more effectively, aiming for better system performance.

Driving Business Performance

From our examples, we assumed your business had 100 energy points allocated to it. For optimal success, undesired output should be as close to 1% as possible, while desired output should approach 100%. To enhance performance, you must channel as much energy as possible toward high-quality, organized work. However, if your system is cluttered with disorder, your efforts may fall short of producing meaningful results.

To truly get your business on track, you need to address the root causes of disorder. Reducing wasted energy or undesired output can significantly boost your success, but the real question is: How do you identify and minimize energy waste?

Identify Undesired Outputs

Many business books and programs emphasize strategies that focus on working harder, but the real solution lies in maintaining a more efficient system. This begins with selecting the right business model. Your business model determines the strategy you'll use to optimize organizational effectiveness. Approaching business from a strategic advantage becomes the single most critical factor for success.

The hard truth is that most business models are inherently chaotic, often stemming from flawed startup strategies. They aren't designed as foundational frameworks, which leads to frustration, burnout, stalled growth, ineffective customer acquisition, and misalignment with organizational culture. I learned this firsthand—only when I applied the laws of physics from *Mindset Mechanics* did my efforts finally produce positive results.

In 2021, I entered the trucking industry. Within 12 months, the business grew into a million-dollar operation, earning recognition for our company culture. But I underestimated that all systems tend toward disorder. As we scaled, signs of disorder emerged, resulting in a major contract loss. We struggled to maintain profit margins, and internal energy became insufficient. With mentorship, coaching, and realignment, we reduced disorder and recalibrated for better performance.

If your business is experiencing wasted energy, address it immediately. Whether it's hiring the right employees, securing better clients, sourcing dependable vendors, upgrading equipment, or seeking expert advice, eliminating energy waste starts with choosing the right business model strategy.

Avoid Undesired Output

Maximizing desired outputs and minimizing undesired outputs begins with understanding how your business system is meant to operate. Energy flows from the inside out. Eliminating undesired outputs starts with aligning the leader's mindset, maintaining physical fitness, and nurturing emotional health. Take a close look at your circle of influence and ensure minimal friction to help build momentum.

High levels of disorder or undesired output indicate that the system is not functioning as designed and that energy isn't integrated properly. In an organization, this signals that the vision and values need realignment. Address energy waste promptly and optimize your business system. If the waste is beyond repair, seek immediate help from an optimization expert. Problems cannot be resolved using the same mindset that created them. Fresh eyes bring a higher energy source, making it easier to identify and eliminate waste and providing the clarity needed for change.

Mastering both art and science is essential for overcoming disorder and chaos in your organization. The *Mindset Mechanics* approach involves stopping undesired outputs, establishing a strong system structure, and implementing continuous maintenance of subsystems to keep everything running smoothly.

---

29. He, J. (2011). Complexity in Adaptive Systems. In: Sammut, C., Webb, G.I. (eds) *Encyclopedia of Machine Learning*. Springer, Boston, MA. https://doi.org/10.1007/978-0-387-30164-8_148
30. Cengel, Y.A and Boles, M. Thermodynamics an Engineering Approach by Yunus A Cengel and Michael A. Boles p. 249
31. Cengel, Yunus A. (2002). *Thermodynamics: An Engineering Approach.* 4th Edition. pg. 249.
32. Hawkin, Stephen A. The Direction of Time (1987), *New Scientist.*

CHAPTER 13: The Roadmap to Drive Values

Let's set the stage: performance in business isn't just about profit, systems, or growth; it's fundamentally about values. Values drive decisions, determine strategy, define culture, and inspire people. Values are at the heart of every organization. The *Roadmap to Drive Values* is a framework designed to guide leaders across all kinds of organizations—whether in a small business, family unit, or international corporation. Values shape and impact change needed for building a high-performance culture.

Positioned in the Business Vehicle Control System (BVCS), the *Roadmap to Drive Values* offers a framework to unlock your organization's potential by aligning what organizations value most—people (see **Figure 29**). This framework aligns what you, as the leader values, team members value, and markets value in an organizational setting.

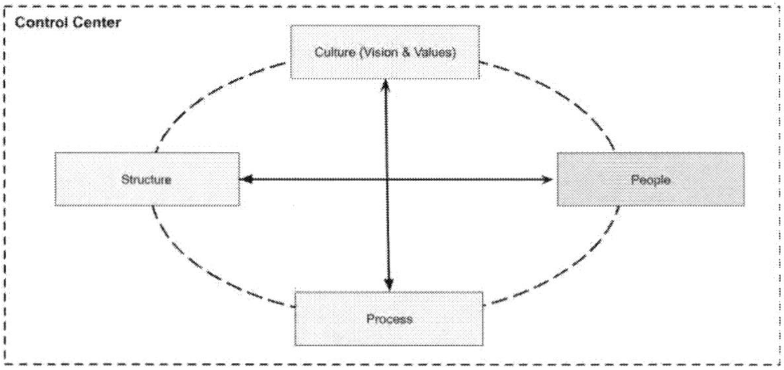

Figure 29. Business Vehicle Control System: People

Have you ever wondered why garbage workers get paid less than CEOs? Why do classroom teachers get paid less than education Administrators? Trash removal and educating our children are essential societal functions, yet they are compensated less than the CEO or Administrator. This framework explains why essential workers are compensated at different levels than decision-makers. Why guarding, nurturing, connecting, and creating change through the organization increased by degrees of value.

Success is Based on Values

Values are determined by their priority of importance. Determining value starts with defining and recognizing acceptable value so you can serve, inspire, and foster positive organizational change. As a vital leader and mechanical operator, how you drive performance in your organization is to remember that value is energy and energy is finite, know the differentiable traits of what is valuable, and know how to find, identify, and extract value from the environment.

The *Roadmap to Drive Values* focuses on organizational and marketplace values that connect the right people into your tribe, manage forces that propel success, and develop external relationships based on a good reputation that drives your mission forward. Central to this roadmap is the understanding that people are your greatest organizational asset. It's about the person you evolve into, the people you bring on board, the customers you serve, and the partners you make. Organizational values aren't just about managing people; they're also about cultivating a shared vision with others, and having that vision resonates deeply with those who will join you on the journey.

The performance of your business vehicle—your organization—depends on perceived values. In the Journal of Business Research, a case study by Jonas Holmqvist and his team entitled "Understanding the Value Process: Value Creation in a Luxury Service Context" sheds light on the different activities of service providers and customers. During the interaction between service providers and customers, they cocreate to determine the activities and processes of value, as shown in **The Value Process Map** in **Figure 30.**

The Value Process Map identifies the five key stages of activities valued by the interaction: value anticipation, value facilitation, value cocreation, value creation, and value learning. This map helps us understand that organizational and marketplace value is between service providers and customers who cocreate value and are separated by two spheres of value: (1) before value cocreation and (2) after value cocreation.

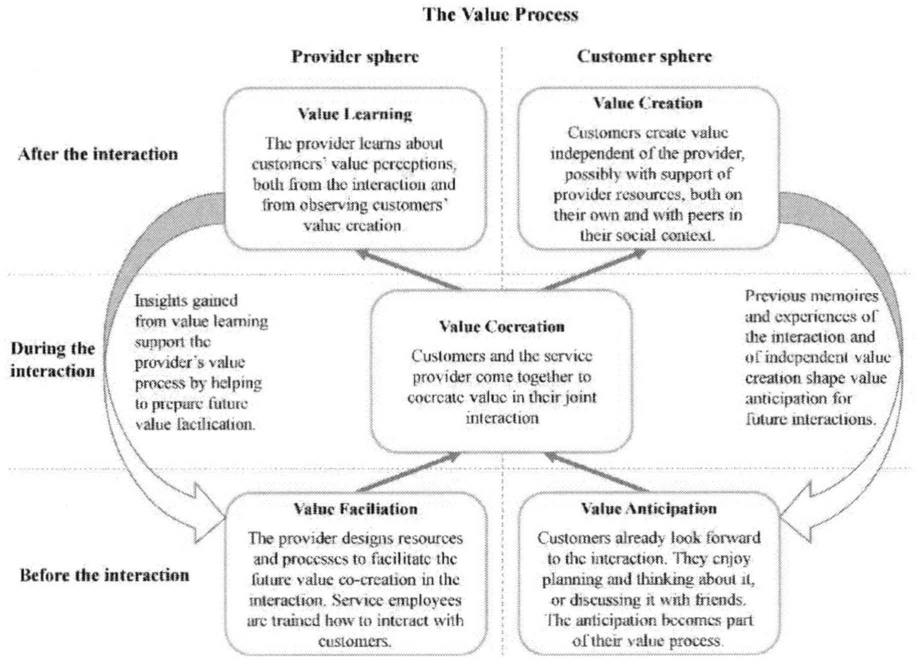

Figure 30. The Value Process Map

1. Before Value Cocreation is Visible or Physical: In this sphere, **Value Anticipation** cocreates with **Value Facilitation** before the customer interacts with the service provider who designs or trains resources and processes to facilitate future work.

2. After Value Cocreation is Invisible or Spiritual: **Value Learning** cocreates with value creation after the customer interacts with the service provider, who learns about perceptions from the interactions and observations of the customer's **Value Creation**.

When values align, they spark interest.

Selling is the process of co-creating value, and it occurs when service providers respond to perceptions shaped by the customer. Success, performance, and wealth are all based on mastering the art of selling to convey your vision, connect with people, and transform ideas into outcomes. Selling is not about convincing others to buy what they don't want. Selling is about discovering unmet needs, revealing the value of your offer, and showing how this value can make a meaningful difference in meeting the needs of their lives. When values align, it sparks interest. Mastering the art of selling values elevates your organization up the value chain, grows your reputation, increases your impact, and establishes your legacy as a leader.

Master Uncovering Value with

Here's where the real transformation begins. The challenge we face isn't simply in showcasing the value we see—it's about helping others recognize and appreciate the values that matter to them. Revealing value is never about us; it's about creating space for others to connect with their own priorities. I value what you value because that's where our shared motivation lies. My role is to support you in evaluating your beliefs to extract, discover value, or create value, not to impose mine.

To foster true alignment, you must blend your values with those of others. When you meet people where they are—giving them what they seek while guiding them toward what they truly need—you cultivate a balance. In physics, this harmony is called **thermodynamic equilibrium,** and it brings a sense of shared purpose and understanding. Alignment elevates all parties and offers insight into seeing new perspectives while experiencing the compassion of being seen and valued. The real value lies not only in what you can provide but in empowering others to think expansively. Give people what they want, but reveal to them what they need. This is the essence of transformative service. Embrace the journey of discovery, and watch it inspire growth in everyone it touches as you encourage others to change.

"People do not care how much you know until they know. how much you care." **Theodore Roosevelt**

If you're genuinely committed to impacting people's lives and building a meaningful legacy, offer more than just a product or service; sell values. This means you deliver a deeper, energetic value that resonates and connects with others. This is the essence of true wealth and prosperity. Embrace this perspective, and you'll notice how it amplifies both your influence and success.

Persuade; Don't Convince

Leaders who lack the skill of persuasion often struggle to create a compelling vision that inspires others to embrace it. Selling is not about convincing someone to agree with you. It's the art of facilitating a value exchange that captures anticipation that compels action. When you persuade, you don't just want agreement—you want genuine alignment, where others see the benefit for themselves and take action.

Convincing might get someone to buy what you offer, but persuasion goes beyond that. Persuasion is about inspiring others to cooperate and cocreate by taking a step to align with their best interests. Persuasion is driven not by what *we think* others need but by uncovering what truly fulfills their needs.

Mastering the art of selling starts with being a great listener. This means engaging with what someone is actually saying—not just to secure a transaction but to unearth the real value hidden behind their words. Listening lets you connect with their experiences. Imagine yourself in their position: *what would you need most?* This shift allows you to ask insightful questions that help reveal what they truly want and, in turn, gives you valuable insights that help you interact, learn, and observe how to support them, what processes they value, anticipation they expect, experiences they've had, and memories they have.

After the interaction, you'll find you have valuable connections or resources to support their need—even if unrelated to your core business offerings—do what's necessary to cocreate value after the interaction. This approach builds trust and goodwill by giving them what they desire before anything else. As your relationship grows, so does your influence to naturally guide others toward the knowledge and skills that can genuinely empower them to exchange their circumstances for values.

Get to the Point

You might believe selling is about being a smooth talker, but listening is the true key to sales. Let me be clear: hearing is not listening. Hearing is passive, and listening is active. Hearing is physical; listening is spiritual. The art of active listening elevates your attention, presence, and connection. It's a component of devotion to others, nurturing, and wonder. Listening is the superpower that elevates your ability to sell.

To master the art of selling, you need to develop exceptional listening skills. This concept is rooted in the essential balance of giving and receiving. Before you can give value, you must receive value and become a receptive listener. Effective communication flows in two ways. In sales, listening is powerful when you're attuned to the interests and needs of your audience.

When you listen, you treat people as intelligent individuals. In return, they see you as knowledgeable. Show genuine respect, and they'll reflect that respect back. Approach sales with a mindset that prioritizes their best interests, and your service is elevated to its highest potential—your clients will value and appreciate you in return. This is the law of reciprocity in numbers theory originating from Swiss mathematician and physicist Leonhard Euler, French mathematician Adrien-Marie Legendre, and proved by German mathematician Carl Friedrich Gauss.

As you can see, selling is backed by physics and mathematics. With confidence, you can listen to the stories of your customers, clients, and team members while you build the skills to translate the details into value. Here is a simple formula: sharing content plus listing for context equals contextual selling. Contextual selling is a sales strategy that shapes seller-buyer engagement using data-driven signals far beyond traditional customer journey mapping. Contextual selling is the future of selling.

Here, sharing their life's content is their motivation—the driving force behind the change or transformation they seek. You bring context to the interaction, aligning resources to find the best path for them to reach that transformation. If you don't have a solution, don't force one;

sometimes listening is the best solution; you can always reconnect if a solution is discovered.

Sell the Payoff, Not the Process

Do you want others to see your offerings as their best solution for transformative change? Show them the payoff your solution brings.

When discussing the benefits of your offer, consider using the word *"potentially"* in your sales process. Using this word is not about creating doubt. It's actually a powerful way to build trust and honesty. Saying, "You could *potentially* feel more fulfilled in your work" or "Imagine *potentially* having more energy" emphasizes transparency and safeguards against overpromising. You always want to under-promise and over-deliver.

Avoid dissecting your solution into small details. Refrain from saying, "It includes this feature and two of those widgets." When it comes to selling, one wants to know the breakdown. They can read the breakdown. What they really want to know is *how it will benefit them*. They need a straightforward path to success, regardless of the effort it may take, and they want the confidence that your solution will deliver for them. Therefore, presenting the benefits clearly and confidently will make people more inclined to buy in, subscribe, or engage wholeheartedly with your offer. Ultimately, your benefits are aligned values that feel like self-love. With this understanding of what draws people to an offering, it's time to turn our attention to the more profound impact of organizational value.

Four Levels of Organizational Value

Now that you understand the balance between needs and values let's delve into the Four Levels of Organizational Value. These levels are essential for any leader aiming to drive meaningful change.

To achieve your desired results, develop a deep understanding of the four levels of value. These levels shape your business focus, organizational talent mix, and customer service approach. If high performance is your goal, your business must function and serve at the highest level appropriate for the required change.

It's also crucial to recognize that value itself is spiritual—it's not material or visible but rather an invisible, high-energetic force. Value is a hidden currency. In this sense, wealth is also spiritual, gained only by becoming a high-energy person, engaging in high-energy activities, and accessing high-energy sources. The Four Levels of Organizational Value are Creating Innovation, Connecting Communication, Nurturing Unification, and Guarding Implementation. Let's start by exploring how these four levels apply to the individual or the Four Levels of Individual Value.

In **Figure 31,** you can see the implementation level is the lowest level, where the individuals doing the work get the least paid. But on the level of creating innovation, individuals are compensated well.

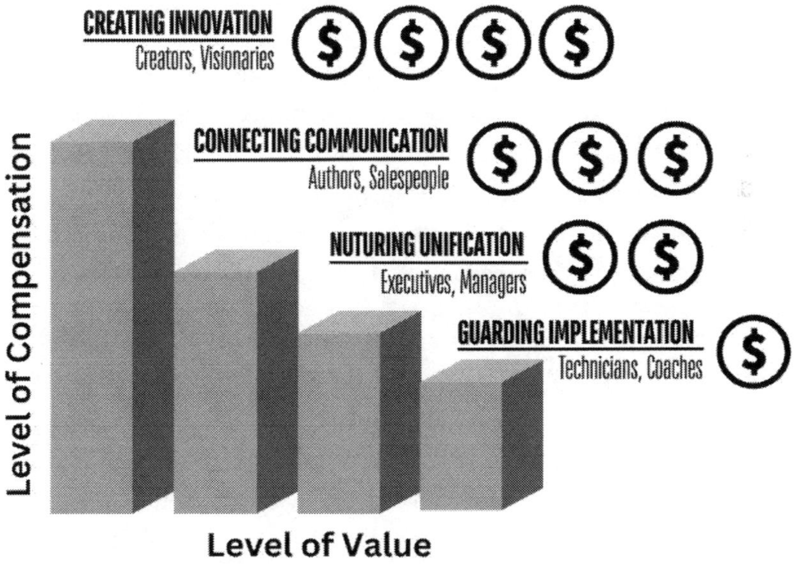

Figure 31. Four Levels of Individual Value

Level 1: Guarding Implementation

Guarding Implementation represents the lowest level of value within an organization. Individuals implementing tasks at this level are essential workers and receive the lowest financial compensation. COVID-19 taught us that being an "essential worker" doesn't mean people are unimportant. It means the role is "generic." Generic things are crowded and have no unique value. Guarding implementors can be replaced, and others can be trained to do the same work. What's happening is the guarding implementor rents out their vacant labor for compensation. The occupancy is filled with an occupation, and the service provider is the occupant who pays for the labor. Risks are low because the service provider trains the guarding implementor and absorbs the risk. Labor pressures are high because the largest pool of implementors exists here. It's low-risk tolerance and high labor pressure that drive compensation down. Guardian implementors are the first line of defense that keeps our societies running. Pay for the working class rarely increases. As you can see, Guardian Implementors are physically integrated as the people resources and processes that facilitate and create value. They are the value. You cannot detach the body from the task at this level.

Consider the hotel industry. Housekeeping staff work tirelessly yet are compensated minimally. Despite being called "essential workers," their pay is disproportionate to their physical contributions. They provide physical resources—labor, strength, and time—to the company and market. And since muscles and labor are finite physical resources, generating high value becomes challenging at this level.

These principles aren't arbitrarily set; the laws of physics govern them. Attempting to achieve boundless, high-impact outcomes through limited physical resources is inherently demanding. According to the laws of thermodynamics, achieving high-quality results with low-quality energy is impossible. This reality often necessitates employing more individuals to meet demands, as physical resources alone cannot yield spiritual results. Impact that lasts draws on unlimited, invisible, universal, and spiritual resources. On the contrary, guardian implementors hold systems grounded.

At this level, the Guarding Implementation organizational value drives essential and foundational tasks and is constrained by the vast amount of physical resources required. This organizational value can be seen in professions that create value: technicians, laborers, and teachers.

Level 2: Nurturing Unification

The next level up is Nurturing Unification. Instead of solely performing laborious tasks, at this level, you're leveraging your leadership and management skills to enhance productivity, earnings, and teams. You manage the people and processes that facilitate and create value. The nurturing unification roles supervise and guide others to achieve results. The risk tolerance decreases for the service provider because, in most cases, the nurturing unifier has paid into their training, resulting in slightly lower labor pressures as competition decreases and tasks become more specialized. Nurturing Unification carries greater spiritual responsibility than implementing tasks. This is why managers often earn more than the individuals they oversee—they scale their efforts by channeling the collective efforts of others toward a shared objective. It's important to note that if guarding implementors are essential workers, based on their degree of association, managers are also essential workers by default.

The Nurturing Unification organizational level of value binds teams together, providing structure and elevating the contributions of others, which supports the individual and organization's ability to grow, strengthen, and manage resources. This organizational value can be seen in the professions that require management of the people or things that create value. True wealth cannot be created at this level of organizational value, but a comfortable living is possible.

Level 3: Connecting Communication

Connecting Communication holds the second-highest level of value in an organization, as it requires unique insight and the ability to connect deeply with others through voice, language, and data. This skill is reserved for those with a higher level of spiritual awareness and often commands greater financial rewards. For instance, an author typically earns more than the person who handles the book's publication. They value their ability to craft and deliver messages that resonate and connect ideas through the spoken word.

To achieve financial success, it's essential to excel in communicating persuasive messages that genuinely connect with people. Most people do not take the time, attention, or investment to get good at communicating and creating valuable results. Mastering your communication skills opens doors to substantial income and the potential for lasting wealth. I am confident that by developing my communication skills, I will earn a far higher income through speaking engagements than I would in a traditional role. The key to wealth creation lies in mastering effective communication. Strong communication skills are indispensable. They allow you to share impactful messages with larger audiences and generate significant wealth.

The Connecting Communication organizational value has the unique power to influence and inspire others on a broad scale, amplifying their impact and creating lasting value. This organizational value can be seen in professions that require voices to create value: salespeople, playwrights, singers, speakers, and authors. This level of organizational value is where wealth creation starts.

Level 4: Creating Innovation

At the pinnacle of value is Creating Innovation. Those who generate groundbreaking ideas are the ones who create the most significant impacts on society and create opportunities. Consider Apple Inc. Only one name likely comes to mind when you think of this company: Steve Jobs. Although he has been gone for over a decade, Jobs is still synonymous with Apple's legacy. Interestingly, he didn't invent the Apple Macintosh; that credit goes to Steve Wozniak. However, when Apple is mentioned, people rarely think of Wozniak or even Tim Cook. Why? Because Wozniak was the implementer—the technical mind behind Apple products—Jobs was the innovator out front. His ideas and his ability to envision a brand that could revolutionize the world left an indelible mark.

Visionary thinking is a higher spiritual endeavor than speaking. If you aim for growth and financial success, "thinking" should be your foundation. Think at a higher level, speak intentionally, communicate real value, and operate with a vision that guides others. The most valuable leaders recognize the need to create space for others to thrive and understand their unique styles, strengths, and limitations.

The Creating Innovation organizational level of value is for top-tier thinkers. The implantation, unification, and communication of innovative ideas that run the world. This value leads and aligns all levels of organizational value. By engaging at this level, leaders are valued as catalysts for change, making their mark and guiding their organizations to reach new heights. This organizational value can be seen in the professions that require ideas and visions to create value. Operating at this level of organizational values receives the highest compensation for wealth.

Organizational Values are Conservative

It's important to note that the markets determined that the highest level of organizational value is creating innovation because there are so few, and the demand is high. This does not mean the other levels provide no value but that their value is integrated into their respective level and driven by saturation. As the organizational strategy changes, the leveraged organizational force of value changes. The leading value for a startup looks different from the organization positioned for growth. Value is based on perspective and needs. A startup company may pay more for guardian implementors than an organization positioned for growth, just as a startup does not value nurturing unification as much as a

The strategy should align with the most appropriate level of value based on the first law of thermodynamics. Therefore, the four levels of organizational value are consistent with energy conservation. Next, Section II helps us understand how organizational value translates into organizational forces.

33. Laura C. Aiken, *Why Resilience is the Ultimate Leadership Skill*, 2023.
34. Adizes, Ichak K. (2023). *Management Corporate Lifecycles.* pg 107.

SECTION II: The Right Organizational Values for High-Performance

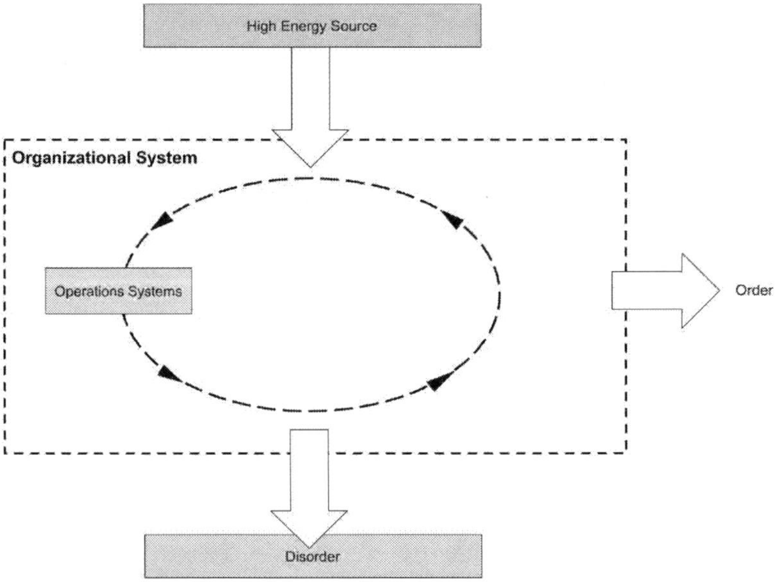

Figure 32. Mindset Mechanics Interface: Operations Systems

CHAPTER 14: Forces of Organizational Value

Every high-performance organization relies on a well-designed operations system—a structure that holds everything together and drives results. Within Mindset Mechanics, the Operations System (OS) is the heart of that structure. In this section ask: *"How to manage teams and lead more effectively?"* The answer guide you through understanding and implementing an organizational structure that's not only stable but adaptable to support a high-performance culture.

Leaders who understand their unique drives can align themselves with the force that best fuels their natural leadership style. However, true success comes from harnessing the full spectrum of strengths within your team, leveraging three additional types of forces to create a complete and cohesive organizational structure. Together, these forces power the resilience and responsiveness of your organization.

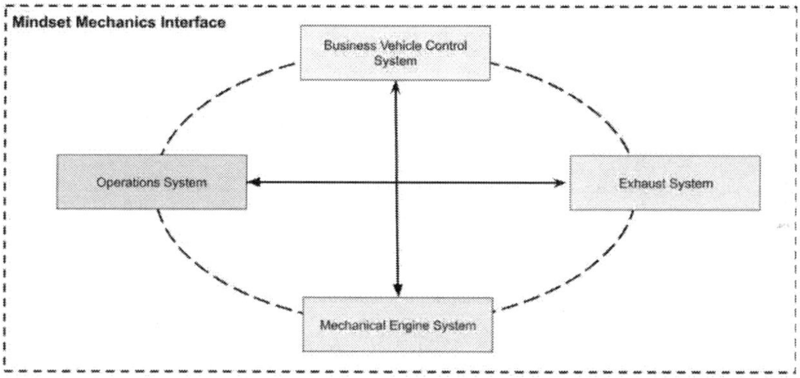

Figure 33. Operations System

In the following pages, we'll explore how high-performance cultures are built on understanding and utilizing these core forces. The Operations System in Mindset Mechanics is designed to provide stability, absorb shocks, and respond to the changing conditions of the business environment, ensuring your vision remains on track. Great leadership emerges from a strong foundation of trust and inspiration. With this system in place, you'll see your organization move forward with confidence, adaptability, and impact.

A Day in the Life of Steve Jobs

In a 2010 interview at the D8 Conference, Steve Jobs offered a rare glimpse into his management philosophy, particularly his approach to fostering collaboration at Apple Inc. Reflecting on Apple's distinct culture, Jobs proudly declared, "Do you know how many committees we have at Apple? Zero." During his time as CEO, he ran Apple with the agility of a startup, avoiding the rigid layers of committee-led decision-making. Instead, he relied on a lean structure: one person led iPhone OS software development, another managed Mac hardware, someone else coordinated global marketing, and yet another oversaw operations. Jobs structured Apple as a dynamic network rather than a hierarchy—the largest startup in the world.

Each week, the leadership team met for three hours, sharing insights on projects, products, and the business landscape. This ongoing dialogue created a culture where trust and autonomy were paramount. Rather than micromanaging, Jobs placed confidence in each team member's ability to deliver, fueling an atmosphere of collaboration and shared accountability. Trust at every level was the cornerstone of Apple's success, driving unity across departments and projects.

Jobs explained, "Every day, I sit down with various teams to brainstorm, solve challenges, and work on developing new products or initiatives." This collaborative energy, he noted, was what allowed Apple to continue pushing boundaries and to empower its talent by giving them the freedom to make meaningful decisions. For organizations, he advised that if they want to attract and retain top talent, they must nurture a culture where ideas take precedence over hierarchy. The most effective ideas must prevail, or valuable team members may lose motivation and leave.

So, how can you integrate the principles Jobs highlighted into your own organization? Start by understanding the unique forces and personalities within your team, empowering each person to contribute their best. Effective organizations are constantly evolving, adapting to change while balancing the needs of each team member with the organization as a whole. This is what I call the Dynamic Forces Model—a universal model that applies to any organization, whether it's a family, a team, or a corporation. It guides organizational change by aligning the right force to lead your operations, driving the results you want.

Four Fundamental Forces of the Marketplace

As we mentioned in the previous chapters, there are Four Fundamental Forces of value seen in individuals, organizations, and the marketplace:

1. The *Strong Nuclear* Force focuses on Guarding Implementation.
2. The *Weak Nuclear* Force pertains to Nurturing Unification.
3. The *Electromagnetic Force* corresponds to Connecting Communication.
4. The *Gravitational Force* is linked to Creating Innovation.

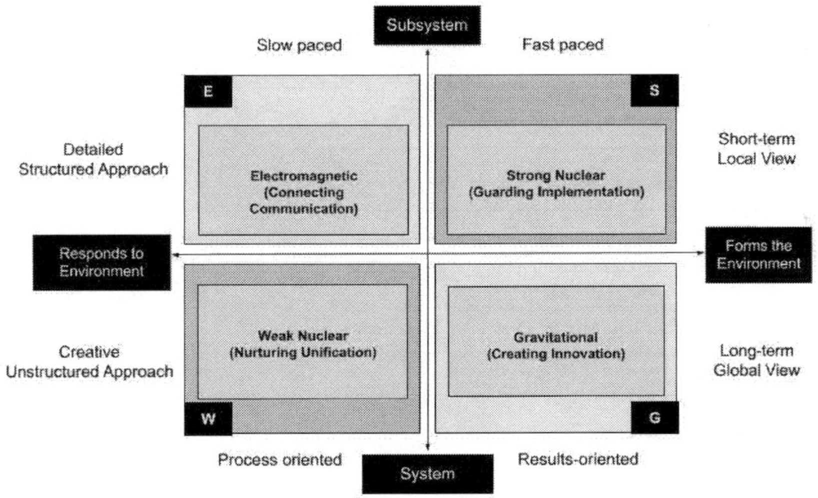

Figure 34. Four Marketplace Values of Mindset Mechanics

Now that you've taken steps to become a leader of change, let's deepen our understanding of these four fundamental forces within your organization. Some may come more naturally to you, while others require intentional focus. Awareness of each force allows you to value the distinct contributions they bring to your team, helping you understand how they complement and strengthen one another.

Organizational Forces of Value are Also Organizational Styles

In this chapter, we explore the dynamics within organizations, equipping leaders with the tools to guide their teams more effectively. To genuinely influence others, it's essential to understand the unique styles that exist within your organization. You'll often see the terms "forces" and "styles" used interchangeably, as they capture a similar essence. Forces reflect a leader's view on the approaches to drive the organization forward, while styles represent the diverse personalities that influence team interactions.

Both forces and styles are integral to moving the organization ahead. They shape every element of the organization, from influencing behavior and defining strategy to determining product offerings, responding to market needs, forming cultural identity, and even highlighting key risks. Recognizing and working with these dynamics, leaders can harness the full potential of their teams and create lasting impact. To help you remember the fundamental force of organizational value, instead of "business swag", I want you to "Think business SWEG."

Strong Nuclear Forces

Strong Nuclear Forces are all about getting things done and driving productivity across the organization. This force assembles the workforce and builds the support structure necessary to execute and achieve results.

The Strong Nuclear Forces serve as the crucial protective and *first line of defense* personality style of organizational success. They are the Guarding Implementers of an organization. They are the true guardians of the enterprise, consistently delivering exceptional results for both customers and clients. They help shape and develop the organization and its culture. Working at an impressive speed, they navigate through various subsystems within the organization, maintaining a sharp focus on the present moment. Their need for information is driven by a short-term perspective, enabling them to tackle immediate tasks effectively. Strong Nuclear personality styles thrive on achieving results, demonstrating unwavering focus with detail. Embracing these attributes is essential for any organization that aims to excel.

Weak Nuclear Forces

Weak Nuclear Forces adapt to changes within the organization, fostering team dynamics, relationships, and company culture. This force nurtures and integrates, ensuring a cohesive and supportive environment that adapts and needs room to grow with the business.

The Weak Nuclear Forces serve as the essential Nurturing Unification personality style within the organization. They are highly adaptive to their surroundings. Individuals of this personality type are inherently creative. Consistently seeking innovative solutions to challenges. They thrive on freedom of expression, so an unstructured approach is ideal for their work because it allows them to operate autonomously. Their decision-making process includes their team. They are relationship-focused and thrive on helping the group. They are cautious and require time to carefully process information before committing to decisions that could have a significant impact on the entire organization. Moreover, they possess a long-term perspective, which is crucial for the overall success and sustainability of the organization. Embracing and supporting their unique working style can lead to remarkable outcomes for all.

Electromagnetic Force

This force is essential for linking and organizing the different parts of the organization. It operates through diverse communication channels, providing stability, structure, and essential administration. The Electromagnetic Force ensures that every part stays informed and aligned. This personality style needs structure and time to process data.

The Electromagnetic Forces serve as the essential Connecting Communicator personality style within the organization. Their ability to respond adeptly to their surroundings sets them apart. With a keen eye for detail, these individuals exhibit a well-organized thought process and structured approach to tasks. They excel in processing, analyzing, and synthesizing substantial volumes of data before arriving at informed decisions. By concentrating on the functioning of subsystems within the organization, they maintain a clear focus on achieving short-term goals and missions. This personality type excels at managing information, data, and documents with impressive orderliness, making them invaluable assets to any team.

Gravitational Forces

Gravitational Forces fuels the organization's potential for growth by encouraging new ideas, challenging norms, and focusing on continuous improvement. Through R&D, product development, and strategic planning, this force drives the innovation that pushes the organization forward.

The Gravitational Forces serve as the Creating Innovators personality style within the organization. This style is a game-changer for any organization, serving as the cornerstone for success. This personality style takes a holistic, long-term perspective, always focusing on future possibilities. Because it is results-oriented, it thrives on efficiency and swift task completion. With its inherently creative personality, this style flourishes in flexible settings. Such an environment empowers individuals to harness resources for innovation, leading to the development of groundbreaking products and services while continuously generating fresh ideas. Embracing the Gravitational Style will undoubtedly propel your organization to new heights.

Putting it all together

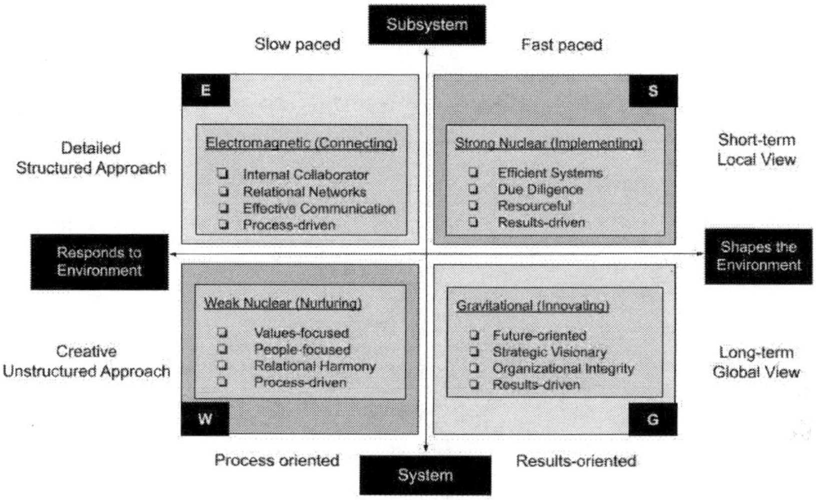

Figure 35. Four Marketplace Values of Mindset Mechanics (Expanded)

There are a few things I want you to notice. First, the right hemisphere is regarded as having masculine traits, while the left hemisphere is considered feminine. These are not gender-specific but spiritual attributes that lend themselves to specific roles and responsibilities. Next, I want to summarize what's occurring on the molecular level of these fundamental forces of nature. The gravitational and electromagnetic forces are based on the law of cause and act universally. These forces are the "subjective why" working behind the scenes with essential particles. Their forces are macro and function as the attraction or the interaction between all things. On the other hand, the strong and weak nuclear forces are based on the law of effect and act physically. These forces are the "objective what" for observable results. Their forces are micro and function to hold things together or change things from the inside out. Finally, notice that each hemisphere has complementary and competing views that create contrast and balance. What is valued by one hemisphere is not valued by the other, yet overlaps exist where value is shared.

Each of us has the capacity to function within any of the four quadrants as needed, but prolonged effort in our less comfortable areas can lead to fatigue and drain our energy. Some forces naturally require more effort and may stretch us beyond our core competencies. As a leader, your aim should be to operate within your most dominant force or personality style about 90 percent of the time, ensuring that both you and your team are working from a place of strength. When team members are empowered to function primarily within their own strengths, it enhances job satisfaction and cultivates a high-performance culture across the organization.

In the next four chapters, we'll take a closer look at each of these forces as they relate to management and personality style, exploring their unique attributes and how to harness them effectively within the organization.

---

35. "Four Forces." *NASA Science,* https://science.nasa.gov/universe/overview/forces/#weak-force. Accessed 2 October 2024.
36. Adizes, Ichak K. (2023). *Management Corporate Lifecycles.*
37. Sisney, Lex. (2021). *Designed to Scale.*
38. Najberg, Adam. "D8: Steve Jobs Onstage: Full-length Video." *The Wall Street Journal,* 7 June 2010, https://www.wsj.com/video/d8-steve-jobs-onstage-full-length-video/ 70F7CC1D-FFBF-4BE0-BFF1-08C300E31E11. Accessed 7 October 2024.
39. Munroe, Myles. *The Spirit of Leadership*. Whitaker House, 2005.

CHAPTER 15: Strong Nuclear Personality Type

Let's dive into the Strong Nuclear Force, a driving force within organizational value that focuses on Guarding Implementation personality styles. This force holds a unique role, aimed at both protecting and executing the fundamental elements that make an organization thrive. Think of it as the cohesive energy that binds your team's efforts into real, tangible results. The Strong Nuclear Force doesn't just support the structure; it builds and reinforces it. By learning how this force operates and how to channel it, you'll be empowered to maximize efficiency and maintain a high-performance environment. This force, when applied well, pushes your organization forward, ensuring goals are met with steadfast precision.

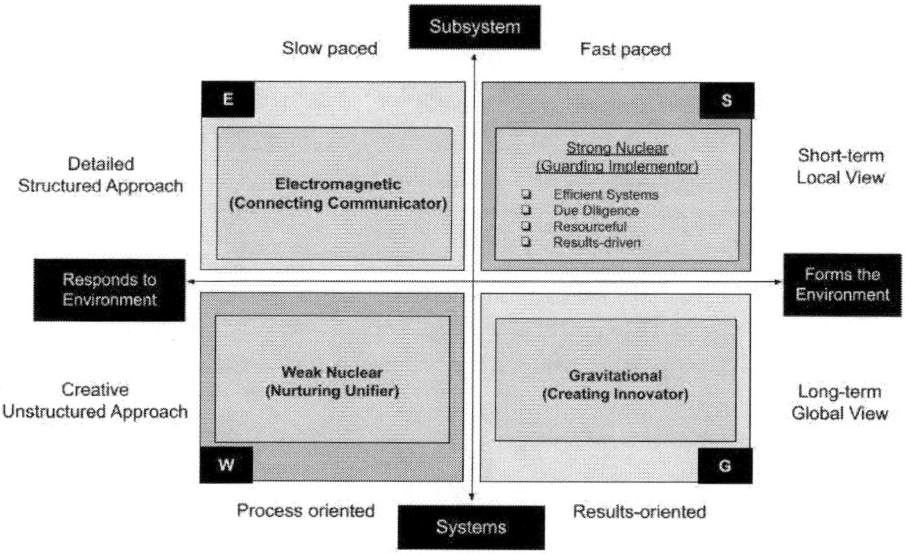

Figure 36. Strong Nuclear Force (Guarding Implementor)

The Strong Nuclear Force represents the backbone of any organization. It's the energy that keeps your processes grounded and efficient, actively driving the completion of short-term objectives. The name fits its purpose—strong, reliable, and foundational, creating a steady flow of progress that others can rely on. This force excels at breaking down complex goals into manageable tasks and is highly efficient and resourceful, getting things done right and on time—every time. If you're looking for a force that can ensure stability and keep the organization running like a well-oiled machine, the Strong Nuclear Force is your answer.

When managing or collaborating with those embodying this personality style, it's essential to understand their nature. Strong Nuclear Forces have a low tolerance for complacency and are generally resistant to change. They prefer clear, consistent routines, finding comfort and productivity in knowing exactly what their daily roles entail. Their natural focus on stability means they excel when allowed to implement tasks with precision, so fostering a structured environment with well-defined expectations will encourage their best work.

Strengths for Strong Nuclear Personalities

Strong Nuclear personality styles excel at absorbing vast amounts of factual data and completing several tasks in their areas of interest. They are incredibly conscientious and hard-working, and everyone trusts them to deliver. They can see the details of tasks and projects, and they enjoy giving them order and structure.

The Strong Nuclear personality is incredibly disciplined in its approach to all aspects of life. They thrive on the adrenaline of meeting challenging targets and deadlines, and they enjoy the intellectual challenge of solving complex practical problems. Competitive environments bring out their best. They are not swayed by spin and charisma. They ask critical questions to ascertain what is true.

Challenges for Strong Nuclear Personalities

Strong Nuclear Forces often thrive in high-pressure environments, where working under last-minute deadlines fuels their productivity. For them, leaving work unfinished simply feels irresponsible. They're driven to complete tasks before allowing themselves to relax. This dedication, however, can make delegation challenging. They may wonder, "Who can do it as well as I can?" which leads them to shoulder the load themselves. Their strong focus on achievement may also make them appear less warm and friendly in a work setting, as they prioritize processes over relational dynamics.

This personality type tends to struggle with building long-term relationships outside of their immediate task focus. Their relentless pursuit of success can create a sense of distance, even from those closest to them, as they are fully absorbed in the tasks at hand. When the Strong Nuclear Force is the dominant style in a leader, they are likely to drive their teams hard, sometimes coming across as demanding and even intolerant of perceived weaknesses. Their intense focus on achieving completing processes can be both their strength and, when unbalanced, a challenge in team dynamics.

Suggested Growth Opportunities for Strong Nuclear Personalities

Strong Nuclear personalities can enhance their leadership effectiveness by learning to connect and empathize with people with different styles. Planning for effective delegation is key. By mentoring others and sharing their knowledge, they can broaden their influence while fostering development within their team.

An important growth area for Strong Nuclear personalities is recognizing that relationships are equally important to task completion. To maintain balance, they should develop healthy routines that include time for work, exercise, friendships, and recreation. Managing impulsive tendencies and tempering their drive can also help prevent burnout, enabling sustainable productivity.

They need to understand that the outcome of a task does not define their self-worth. As they focus on building empathy, they'll find that "good enough" is often just right and that delegation is a critical part of impactful leadership. Additionally, planning to delegate effectively and mentoring others will create a more adaptable and resilient team. Strong Nuclear personalities should remember to relax their expectations—those they're developing may never approach tasks in the exact same way, but their contributions can still be invaluable to the organization's success.

Response to Stress for Strong Nuclear Personalities

Under moderate stress, this force can become a relentless taskmaster, driving themselves and others to the limit, which strains the team. They rely on adrenaline and minimal sleep. To disconnect from pressure, they might seek fleeting escapes. Their focus shifts to controlling immediate details, often withdrawing from others until they regain control.

Under extreme stress, failure leads to withdrawal and the tendency to catastrophize. They may internalize the failure, letting it shape their self-view. If they fall short of an objective, their communication often becomes negative, casting a bleak outlook for the future.

Characteristics of a Healthy Strong Nuclear Personality

They possess a remarkable ability to absorb, process, and interpret data to shape the environment. With a love of life and a relentless drive to push physical boundaries, they communicate clearly and logically, setting clear expectations for others. They thrive when free to apply their skills and expertise to complex challenges, bringing a practical, present-focused, and competent approach. By integrating intricate systems and human dynamics, they create order and safety to meet shared objectives.

Staying Healthy for Strong Nuclear Personalities

To bring order to their internal world, they must carve out time away from external demands, dedicating moments to be present, competitive, and adventurous with loved ones. Scheduling also allows them to organize and bring discipline to their physical surroundings.

They thrive in varied work that highlights their competence across different settings and seek clear success criteria and objectives for every area of their lives. Opportunities to win and earn recognition fuel their drive. To recharge, they need regular time to engage with the external world—often through exercise or friendly competition.

Causes of Unhealthiness for Strong Nuclear Personalities

They dislike being bogged down by routine tasks that don't challenge their skills. They can become frustrated when others don't share their hard work ethic or demonstrate teamwork. When project demands become overwhelming, they feel mentally cluttered and disorganized. If they feel unorganized, they struggle to complete tasks, which leads to the feeling that they could be perceived as being incompetent, which can lead to a fear of failure. Their sense of invincibility sometimes drives them to pursue adrenaline recklessly.

Without time for themselves, they mentally get stuck on an endless task treadmill, feeling as though they're completing nothing well. They may feel overwhelmed by others' needs and the relational demands on their time, leading to frustration when they feel unable to be the spouse, parent, boss, or friend they aspire to be.

This personality style thrives in sales and production roles.

CHAPTER 16: Weak Nuclear Personality Type

To build a thriving organization, understanding the different types of energies that drive it forward is essential. Let's look at the Weak Nuclear Force, which belongs to the Nurturing Unification personality type and plays a key role in supporting the entire organizational system. This force is sensitive to environmental changes and works across the system, harmonizing relationships and sustaining a unified vision. By aligning with this force, you gain a valuable unifier—one that cultivates strong, high-performing teams and fosters a culture where all members feel included and supported.

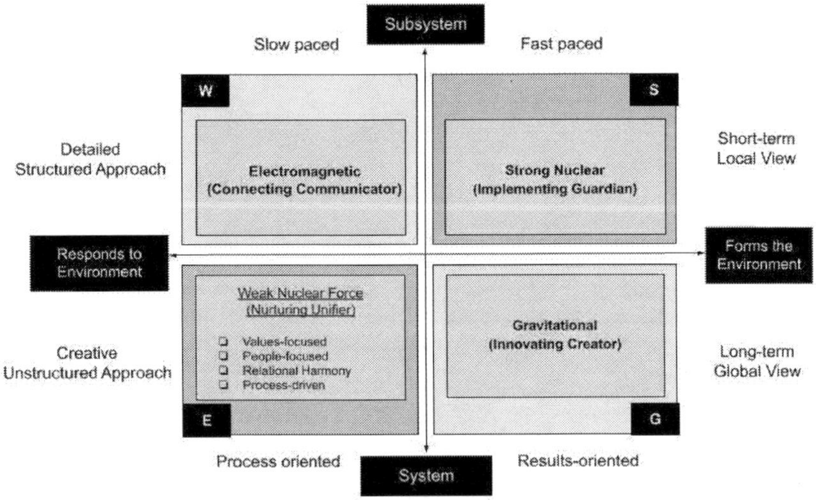

Figure 37. Weak Nuclear Force (Nurturing Unifier)

The Weak Nuclear Force lives up to its name, embodying a steady, nurturing energy that serves as the compassionate backbone of the organization. Its role extends beyond task completion to encompass genuine concern for people and relationships, anchoring the organization in its core values. It is this commitment to the welfare of others that ultimately drives sustainable success.

This influential force focuses on meeting long-term goals through a well-thought-through approach. It guides change, ensuring all members understand and align with new initiatives. Known for refining processes and supporting organizational values, this force also excels in discovering creative solutions to everyday challenges. Leaning into the strengths of this force will help you nurture an inclusive, collaborative environment where team members feel valued and inspired.

If you manage individuals with this force, it's essential to recognize their high standards and commitment to inclusivity. These qualities create a foundation for collaboration, where each person contributes meaningfully and feels engaged in the organization's shared mission.

Strengths of Weak Nuclear Personalities

People with a dominant Weak Nuclear Force personality are deeply selfless and find real joy in the success of others. Highly relational and committed, they build strong bonds with those closest to them, always rooted in the present and attuned to the practical needs of those around them.

They excel at smoothing the wheels of relationships within families, teams, and organizations, embodying immense integrity and a natural drive to serve. Known for their reliability, they thrive in roles where they can fulfill the needs of those they feel called to support. These individuals are not only competent and conscientious but also hardworking and trusted mediators, valued for their sound judgment and strong character.

Challenges of Weak Nuclear Personalities

Weak Nuclear personalities often have perfectionist tendencies and tend to be their own toughest critics, constantly questioning if they could have done more. They're internal idealists, regularly reflecting on their worth and wondering if they're truly helping those around them. These are the reluctant leaders who prefer to dive into hands-on work rather than lead from the front.

They can struggle with keeping projects on time and within

budget, and strategic thinking and project management may feel challenging due to their focus on day-to-day tasks. Immediate demands can easily consume them, making it hard to prioritize recharging. In a combative team setting, they may find it difficult to assert their own needs or to challenge others when necessary. Sleep becomes one of the few places where they can genuinely rest and rejuvenate.

Suggested Growth Opportunities for Weak Nuclear Personalities

Weak Nuclear personalities benefit from establishing daily and weekly routines to recharge and tend to their own needs. Recognizing their idealistic nature, they should be mindful of setting themselves up for potential disappointment. It's essential to understand that they aren't responsible for caring for everyone, and others are there to share the load.

Acknowledging their own limitations is key. Setting realistic goals not only helps them but also strengthens their support for others. Accepting that they can't "rescue" everyone allows them to refocus their efforts effectively. Trusted external voices are invaluable for keeping them grounded, as self-criticism can sometimes become a negative spiral.

They must learn to accept compliments and see themselves as the skilled professionals others recognize them to be. Carving out personal time is essential—creating a balanced routine that includes friends, rest, and recreation. Most importantly, they should feel free to savor the present moment with loved ones without guilt over unfinished tasks.

Response to Stress for Weak Nuclear Personalities

Under moderate stress, Weak Nuclear personalities find comfort in creating order, like planning meals, baking, or preparing gifts well ahead of time. They often question their worth and wonder if they're doing enough for others, holding themselves to unrealistically high standards. This drive to meet everyone's needs can lead to frustration, neglect of personal time, physical exhaustion, and eventually illness. Fatigue and self-criticism can draw them into a negative cycle, risking deeper burnout or depression.

Under extreme stress, they may catastrophize the future, distance themselves from loved ones, and internalize feelings of failure. Highly critical of those closest to them, they wonder, "Why doesn't anyone else see these needs?" They question their own abilities, spiraling downward as they process feelings of frustration. Afterward, when the anger fades, they're often left wondering why it surfaced, reinforcing a sense of perceived failure.

Characteristics of a Healthy Weak Nuclear Personality

Weak Nuclear personalities have an impressive ability to store vast amounts of information about people and causes they deeply care about. Their commitment to providing practical care and advice for those around them is unwavering, as they are fully present, competent, and attentive to the practical, emotional, and relational needs of others.

With a strong drive for integrity, depth, and alignment between principles and values, they remain committed not only to people but also to specific places and communities. The connection between their identity and actions is essential, as they seek genuine alignment in all they do. Their natural "early warning radar" allows them to sense inauthenticity in others, giving them a keen ability to recognize when values don't align.

Staying Healthy for Weak Nuclear Personalities

Weak Nuclear personalities must carefully manage their relationship load to maintain balance. They benefit from spending regular time away from their physical environment and the responsibilities tied to it, creating space to recharge. Scheduling time to tidy and bring order to their surroundings also helps them stay grounded.

Deep friendships are essential for them, as these relationships allow them to be fully themselves. They should set aside time to nurture these connections, care for their physical and emotional needs, and spend quality time with loved ones. Without mindful allocation, others may pull them in various directions, consuming their energy.

Their task world needs to align with their core values, purpose, and life vision. Close, trusted friends who can speak truthfully are invaluable, helping them avoid overcommitment and potential burnout.

Causes of Unhealthiness for Weak Nuclear Personalities

When overwhelmed by the demands of those needing their care, Weak Nuclear personalities often experience feelings of inadequacy. They struggle to "switch off" their inclination to care, leading them to push forward relentlessly until they reach exhaustion. They can become trapped in a cycle of tasks and people-focused responsibilities, neglecting self-care and spiraling into burnout.

This group finds it easy to feel as though they've failed when faced with disorganization or chaos around them, interpreting it as a personal shortcoming. They may neglect their physical and emotional needs and lose time for deep, fulfilling relationships. Witnessing the injustices or challenges of the world can intensify their guilt, driving them even harder until illness forces them to pause. Managing these tendencies is essential for sustaining their well-being and effectiveness.

This personality style thrives in management and operations roles that involve leading or engaging with people and teams.

CHAPTER 17: Electromagnetic Personality Type

Now let's explore the Electromagnetic Force, a key player in the Connecting Communicator personality type and organizational value. This force is vital for bridging gaps, fostering connection, and creating clarity within teams and organizations. Like the unseen yet powerful electromagnetic waves that bring television to life, this force powers communication in an organization. It illuminates information and makes data understandable, allowing people to connect seamlessly.

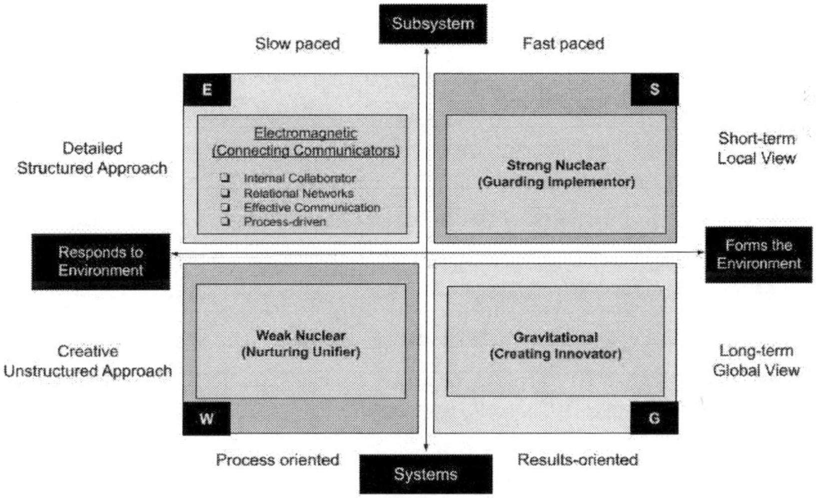

Figure 38. Electromagnetic Force (Connecting Communicator)

The Electromagnetic Force embodies the art of communication and connection. As its name suggests, this force is magnetic and energetic and responsively influences others with messages that resonate deeply within its range. Great communicators generate a magnetic field around them, naturally attracting or repelling others. This power lies in their immediate circle of influence, and they work effectively within their networks to create meaningful relationships.

Managing or working alongside those with a strong Electromagnetic Force requires an understanding of their audience-centered approach. Like charges that attract or repel based on their polarity, these communicators adjust their style to suit the needs of their audience, ensuring that their message aligns with and resonates in the environment around them. In organizations, Electromagnetic personalities adapt communication styles to foster clear, open channels within teams, ultimately ensuring that connection remains at the forefront of every exchange.

Strengths of Electromagnetic Personalities

The Electromagnetic Force is ideally suited for Connecting Communicator personalities in any organization. These individuals have a unique ability to envision an immediate future where people can thrive and unlock their full potential. With a natural charisma and a wellspring of energy for new ideas and relationships, they are resilient, always bouncing back and re-engaging with the world around them.

Responding intuitively to their environment, Electromagnetic personalities assess the bigger picture, drawing connections across the organization. They thrive on order, structure, communication, and documentation. When they genuinely believe in a cause, they ignite enthusiasm in others, becoming catalysts for engagement and inspiration. Their talent and charisma empower them to connect seamlessly, rallying others behind their vision and creating a powerful, united force within the organization.

Challenges of Electromagnetic Personalities

Electromagnetic personalities often find it challenging to harness their talent with consistent discipline, especially as the initial excitement fades. Maintaining focus and structured commitment in their external communication can be a struggle, as their enthusiasm and passion sometimes lead them to overpromise and underdeliver. They tend to take critiques personally, making it easy for them to feel discouraged. For these personalities, a solid framework for follow-through is essential to match their enthusiasm with reliable outcomes.

Suggested Growth Opportunities for Electromagnetic Personalities

Electromagnetic personalities can benefit greatly from practicing discretion and discipline when sharing ideas because they are data-driven. To achieve remarkable results, they should focus on aligning their natural talent with consistent discipline in one specific area. Embracing the challenge of pushing others, even in uncomfortable situations, helps them avoid passive-aggressive tendencies. Commit to mentoring others in their areas of expertise can foster growth, build patience, and deepen their impact. Although they often seek quick wins, guiding others over time will lead to more sustained success and fulfillment.

Response to Stress of Electromagnetic Personalities

Under moderate stress, Electromagnetic personalities often generate a flurry of ideas without giving them the time to develop fully. They need space to thoughtfully plan and organize their insights, data, and information. While they may receive positive reinforcement from authority figures, they can exhibit passive-aggressive behavior toward team members or clients who aren't fully aligned with their vision. Constant changes can exhaust them, as they require a level of stability to function optimally.

Under extreme stress, they may withdraw from others and become overly self-critical, frequently overanalyzing perceived mistakes. This can lead to sensory escapism, such as overeating, overdrinking, or other forms of distraction. It's essential for Electromagnetic personalities to be aware of these tendencies and manage them to stay balanced and effective.

Characteristics of Healthy Electromagnetic Personalities

Electromagnetic personalities are driven by exploring future possibilities and dreaming alongside others. With engaging storytelling and their natural charisma, they inspire people and connect them to new horizons. Passionate about helping others grasp information, make personal breakthroughs, and reach their potential, they are deeply motivated by seeing people grow. They aim to live fully integrated lives, consistently working toward a compelling vision of the future with the people they love and care for, making each step of the journey meaningful.

Staying Healthy as an Electromagnetic Personality

To unlock their best creativity, Electromagnetic personalities need safe spaces where future visions can be freely shared and ideas remain flexible. Brainstorming out loud with trusted others allows their creativity to flow. Planning repetitive, detail-oriented tasks is also essential so they don't get bogged down in minutiae. Above all, they need genuine belief in what they're doing; this conviction fuels their ability to passionately sell a compelling vision of the future.

Causes of Unhealthiness of Electromagnetic Personality

Without the freedom to dream out loud, communicate and connect the details with trusted individuals in a safe space, Electromagnetic personalities may struggle. When fear of failure sets in, they can become reactive, jumping from one idea to the next and ultimately exhausting their leadership credibility. In these moments, they may direct criticism personally, becoming defensive and exhibiting passive-aggressive behaviors that strain relationships.

This personality style thrives in data-driven roles in communications, accounting, quality assurance, and database management.

CHAPTER 18: Gravitational Personality Type

In this section, we explore the Gravitational Force, a key player within the Creating Innovation value in an organization. This force is unique in its ability to shape the environment, with a keen focus on the entire system. Let's take a closer look at the Gravitational Force personality type, examining how it interacts with other organizational forces and styles and how to manage this type effectively.

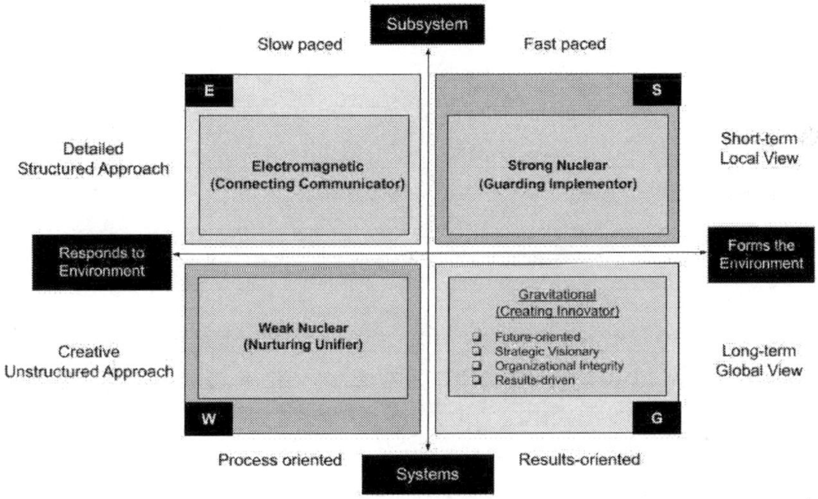

Figure 39. Gravitational Force (Creating Innovator)

The Gravitational Force is the essential element that shapes the organization from within. Operating on two fronts—creation and innovation—this force forms the foundation of a vision-driven culture. The concept of "gravitational force" aptly captures the essence of those who provide strategic vision and long-term guidance. This force has the power to shape the trajectory of an organization by actively creating and refining long-term goals. While their drive for immediate results ensures they are efficient and resourceful, the gravitational force is constantly oriented toward future-focused tasks, ensuring that every move contributes to the bigger picture.

If you manage or work alongside individuals who embody the

Gravitational Force, it's crucial to appreciate the weight of their ideas and the validation they seek for their vision. These thought leaders excel in roles requiring strategic foresight and are often the first to spot industry trends and forecast market shifts. Embracing and supporting this perspective can be the key to fostering meaningful innovation and driving long-term success.

Strengths for Gravitational Personalities

Gravitational personalities are powerful, visionary thinkers who excel at devising fresh, innovative approaches to longstanding challenges. Their deep inner world keeps them grounded, making them less susceptible to fleeting trends or popular ideologies. Instead, they think beyond the conventional, challenging mainstream beliefs and asking profound, thought-provoking questions that encourage teams to uncover deeper truths. This personality style has an innate ability to see change before trends.

With a commitment to lifelong learning, Gravitational personalities are always exploring new areas of study and growth. They are highly relational, deeply loyal to those within their circle, and possess a strong sense of integrity and humility. Willing to support their team in any capacity, they bring an adaptable spirit, making them exceptional coaches and mentors. They hold themselves—and those they train—to high standards, always focused on helping others reach their potential.

Challenges for Gravitational Personalities

Gravitational personalities often operate with a limited need for social interaction, which can make them seem distant or hard to read. Their deep thinking and analysis typically happen internally, and they may engage with the team only after they've crafted a solution. When they do communicate, their initial remarks can sometimes come off as critical, with their true intent not always conveyed in their first expression.

These individuals are continually questioning if they know enough to be genuine experts in their field, driven by a desire for thoroughness and accuracy. They often have strong people-pleasing tendencies, making it hard for them to say "no" and leaving them at risk of physical exhaustion or feeling like they've let others down. These tendencies also cause them to need others to love their ideas. Disagreement or an inability to see the vision can frustrate this personality type and be perceived as a personal rejection because this personality type is closely identifies with their creativity and ability to innovate. Additionally, they may find it challenging to articulate their inner thoughts in ways that resonate clearly with others. Building thought leadership through writing or recording helps flush out creative ideas and crafts more compelling, polish messages for sharing.

Suggested Growth Opportunities for Gravitational Personalities

For growth and development, Gravitational personalities should aim to share their ideas and emotions with those closest to them. By inviting others into their thought process and regularly updating their team on current ideas and goals, they can foster a deeper connection and build trust. Mentoring others not only helps them share their expertise but also enables them to articulate and refine their own insights and skills, making their unconscious competence more accessible to the team.

It's valuable for Gravitational personalities to intentionally affirm and recognize the unique contributions of others, especially those with different perspectives, which enriches collaboration and strengthens team dynamics. They should also prioritize time in their schedules to step away from social interactions, allowing them to recharge and stay aligned with their larger vision. By recognizing their people-pleasing tendencies, they can develop a filtering system to avoid immediate responses, ensuring they thoughtfully consider each commitment. Embracing how much they know and the fulfillment they gain from mentoring can be transformative. Gravitational personalities should take on the leadership challenge, bringing their visions into the world to help others, rather than settling solely for a supportive role.

Response to Stress for Gravitational Personalities

Under Moderate Stress: Gravitational personalities may lose patience with others and undervalue their contributions. Without enough alone time, they struggle to process new ideas and can become distant, weakening key relationships within their team. Overextension often happens as they find it difficult to say "no," leading to self-criticism and occasional tearfulness when stressed. They may also overanalyze their interactions, second-guessing each engagement.

Under Extreme Stress: Doubts surface, and they start questioning their own worth and leadership abilities. With an unclear sense of direction, they become increasingly self-critical, and there's a temptation to escape stress through retail therapy or indulgent distractions. By understanding these tendencies, Gravitational personalities can take proactive steps to manage stress and preserve their big-picture vision.

Characteristics of a Healthy Gravitational Personality

Gravitational Personalities possess a keen intellect and a passion for solving complex problems, driven by a relentless desire to question the status quo and expand the horizons of thought and innovation. Their analytical skills allow them to evaluate others' contributions effectively and ask penetrating questions that reach the core of a problem or opportunity. Fueled by a rich inner world, they constantly seek to understand the broader universe around them.

They have a unique ability to envision inspiring future possibilities that empower themselves and those around them to reach their fullest potential. With a penchant for deep thinking, they enjoy tackling life's big questions on topics like faith, purpose, and the cosmos. In relationships, they are both generous and long-term oriented, making them invaluable friends who invest deeply in the growth and success of those closest to them.

Staying Healthy as a Gravitational Personality

Gravitational Personalities should prioritize regular, dedicated time to nourish their inner world away from people and day-to-day tasks, allowing space to gain a broader perspective. Strategic time investment is essential, as they often have limited capacity to engage fully—physically, emotionally, and intellectually—in every interaction. Scheduling time with close friends or communities that enjoy exploring big ideas and philosophical discussions is invaluable for them.

To stay aligned with their vision, they need intentional breaks from daily demands to re-center and view the big picture. This includes reserving creative time to read, study, explore emerging trends, and envision future possibilities. Investing in deepening key relationships also supports their growth, grounding them as they expand their influence and pursue their long-term goals.

Causes of Unhealthiness in Gravitational Personalities

Gravitational Personalities can feel trapped by others' emotional needs, finding it hard to detach and regain their intellectual space for dreaming and envisioning the future. They may feel stuck in repetitive tasks that leave little room for big-picture thinking, or begin doubting their own competency if their abilities are questioned.

When overwhelmed by practical details, they may double down on perfectionist tendencies, working harder but losing productivity. Without dedicated time for personal growth, recreation, and nurturing their closest relationships, they risk burnout. They can also fall into the trap of believing they must be "perfect" before sharing their knowledge, stalling their potential to contribute meaningfully. Balancing their inner drive with scheduled time for growth and connection is essential to staying aligned with their purpose. This personality style thrives in roles that require big thinking, such as strategy, marketing, and research and development.

SECTION III: The Right Business Strategy Makes All the Difference

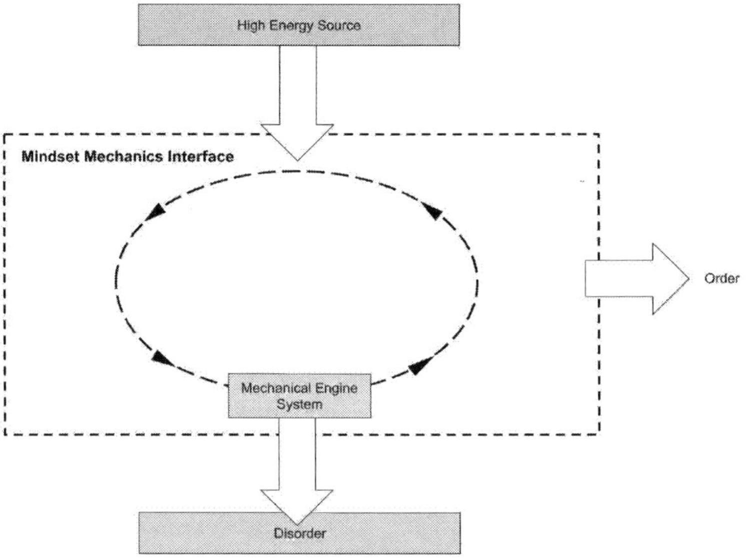

Figure 40. Mindset Mechanics Interface: Mechanical Engine System

CHAPTER 19: The Mechanical Master Plan

There is a master plan for strategy, and it's found here in the *Mechanical Engine System (MES)* (see **Figure 41**). The Mechanical Engine System is the strategic powerhouse for achieving success through synchronized decision-making and implementation designed to help you choose the right strategy that aligns with your long-term vision and allows you to stay committed to and execute.

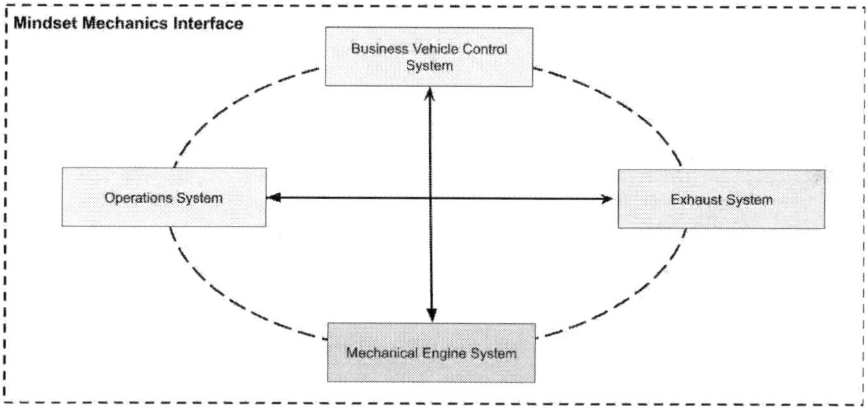

Figure 41. The Mechanical Engine System

The strategy has two main parts: the *Mechanical Fuel System* (MFS), which is focused on adaptation triggers, asks: "Where am I?" and the *Mechanical Systems Lifecycle (MSL)*, which is focused on developmental triggers, asks: *"What's next?"* When these two parts are aligned, they generate an efficient, synchronized strategy that drives real results.

The *Mechanical Systems Lifecycle* involves identifying patterns and stages. By aligning the lifecycle strategies in the correct sequence against the *Mechanical Fuel System*, you enhance the likelihood of achieving your goals. Meanwhile, the *Mechanical Fuel System* consists of four strategic elements critical to business survival: business models, products, markets, and execution. Mastering these four areas allows you to harness energy from the environment now and in the future, ensuring a resilient, growth-oriented approach.

Applying the *Mechanical Engine System (MES)* helped me scale my business to seven figures, and I now guide clients in using this framework to reach their potential. Before I discovered the MES, business was a passion but not yet a lucrative venture. I often felt overworked and underpaid. By using this system and adopting a growth-focused mindset, the principles of physics naturally supported my business expansion.

Because this system emphasizes strategic decision-making and implementation, I can work *on* my business rather than getting trapped *in* it. This approach is essential for building scalable systems, attracting a committed team, and forming a sustainable network of suppliers and partners. I aim to help you shift from entrapment to delivering freedom-driven processes. I'm sharing the details of this system with you because MES is effective for any business, regardless of background, and is accessible to anyone willing to implement it.

The success my clients and I have achieved through Mindset Mechanics isn't magic but real results aligned with universal laws. I'm not offering theory but showing you real principles grounded in mechanical systems and thermodynamics, backed by practical business experience. A thriving system aligns with these principles; one that doesn't will likely struggle. These are the same elements I've used to build my own business.

If you're ready to fuel your inner drive and achieve your business dreams, you must apply the MES strategically. This involves adapting through the *Mechanical Systems Lifecycle* stages—from seedling to sprout, sapling, and aging before dying out. It also requires implementing the four core components of the *Mechanical Fuel System*: defining your business model, refining your products, identifying the right markets, and executing effectively. Adapting these elements to align with the Mechanical Systems Lifecycle is key to creating a thriving, adaptive business.

What is Lifecycle Strategy?

Lifecycles and adaptation are visible across nature, illustrating that growth and survival rely on optimizing the lifecycle or slowing the effects of adaptation. Consider trees: they progress from seed to sprout, then seedling, reaching adult maturity before finally aging and decomposing as logs. Because trees are renewable, their lifecycle restarts through planting or natural regeneration. Similarly, tadpoles hatch in water, transforming rapidly if their habitat begins to dry out. While in stable environments, they mature more slowly. The adult frog can survive on land and eat differently, allowing both stages to thrive without competing for resources—a concept that also applies to caterpillars becoming butterflies.

Human lifecycles follow similar stages, adapting at each phase. Just as we wouldn't expect an adult to act like an infant in a business setting, people develop cognitively and socially to align with their changing environments. An adult who appears fully mature might struggle in an unfamiliar setting, requiring different skills or behaviors, and needing time to adapt. This concept of adaptation across lifecycles is normal and crucial for organizational success, as well.

In his influential Harvard Business Review article, *Strategy*, Michael E. Porter presents three essential principles for defining and executing a successful strategy:

1. **Creating a unique and valuable position** by engaging in a distinct set of activities.
2. **Making trade-offs** in competition, particularly in deciding what *not* to do.
3. **Creating "fit" among a company's activities** to ensure they complement and reinforce each other.

Porter's framework highlights that strategy is about deliberate choice, alignment, and adaptability rather than simply competing to be the best. Strategy also hinges on execution within three core elements: *habitat*, *adaptation*, and *lifecycle*.

- **Habitat**: the specific environment in which an entity—such as an organization—operates.
- **Adaptation**: the set of traits or adjustments that enable an entity to survive and thrive within its habitat.

- **Lifecycle**: a process encompassing distinct stages, from birth and growth to aging and eventual decline.

An organization's growth and sustainability rely on its ability to adapt throughout each stage of its lifecycle. Just as organisms transition through life stages, organizations experience phases in their development, requiring continuous adaptation to remain viable. We represent the lifecycle phases as *organizational development* and adaptation as *organizational stability*, as seen below in **Figure 42**. This model provides a roadmap for leaders to navigate growth and change strategically, allowing their organizations to evolve alongside their unique environment.

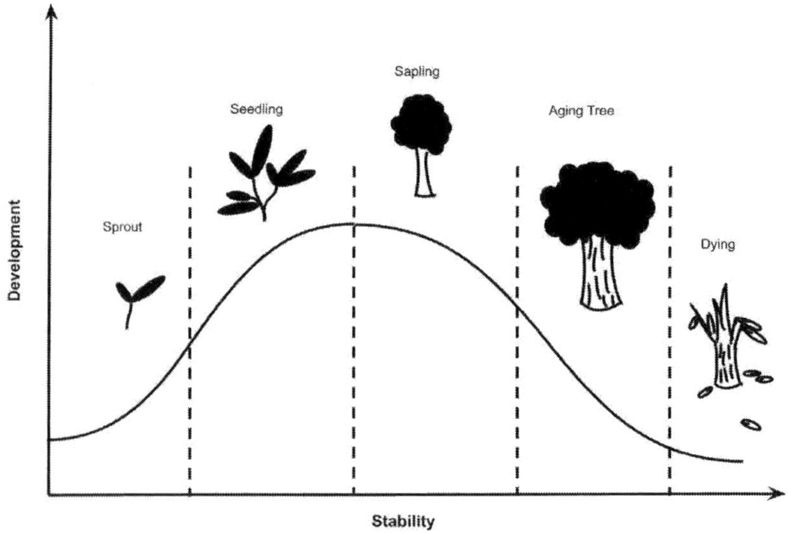

Figure 42. Lifecycle Development

Mechanical Fuel System

The *Mechanical Fuel System* (refer to **Figure 43**) maps the adaptation during decision-making and implementation processes. While the organizational energy needed for your business system is readily available in the environment, tapping into it requires understanding each component's function and triggers within the framework—fuel types, pumps, lines, and filters.

In this model:

- **Fuel** symbolizes the business models that drive organizational energy.
- **Fuel pumps** represent business products, the engines of revenue and growth.
- **Fuel lines** convey energy to business markets, targeting specific customer bases.
- **Fuel filters** reflect business execution, ensuring the strategy flows efficiently without obstacles.

The effectiveness of the *Mechanical Fuel System* lies in correctly adapting and executing each component strategy within the *Mechanical Systems Lifecycle*. This alignment ensures a smooth transition of energy from the environment to peak organizational performance, advancing your organization through each lifecycle stage with purpose and precision.

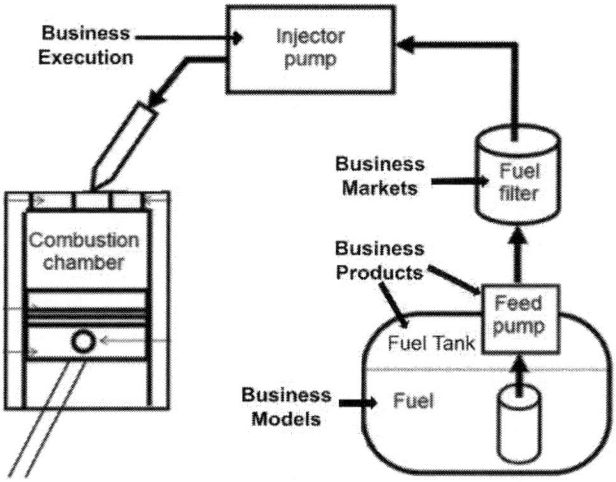

Figure 43. Mechanical Fuel System Flow Map Leading to the Mechanical Engine System.

Organizational adaptation involves recognizing the specific lifecycle stage of the business and applying the right strategies accordingly. This is why some companies thrive while others fail—they miss adapting the appropriate strategies at the right time. In the *Mechanical Fuel System*, we identify four key lifecycle strategies essential to thriving at each stage:

1. **Business Models** - Define the energy or resources extracted from the environment to fuel growth.
2. **Business Markets** - Focus on targeted customer bases for sales efforts.
3. **Business Products** - Offer assets developed for sale, aligned with customer needs.
4. **Business Execution** - Execute these strategies with precision to meet goals.

Successfully applying these strategies as your business progresses through its lifecycle will create alignment and drive sustainable growth. By intentionally managing these strategies, you harness the Mechanical Systems Lifecycle to lead your organization to long-term success.

Putting it All Together

In business strategy, ensuring that all four components of the *Mechanical Fuel System*—business model, market, product, and execution lifecycles—are seamlessly aligned with the *Mechanical Systems Lifecycle* is not merely advantageous; it is essential. This integration is the foundation for sustained growth and success, as each lifecycle component has distinct behaviors at every stage of development, much like the predictable milestones in early childhood development.

As illustrated in **Figure 44**, every phase of the *Mechanical Systems Lifecycle* carries specific, expected characteristics tied to these strategic components. For example, an open business model lifecycle operates in a well-defined and predictable manner, while the prototype in the product lifecycle adheres to established patterns. Likewise, the Fast Lane market lifecycles and the sprout phase of organizational development demonstrate unique traits essential for advancement and supported by the Gravitational Force and Creating Innovation personality styles to maximize results. Overlooking this alignment could stifle growth and limit your business's effectiveness. However, by maintaining synergy between each element, you can elevate your strategic approach, drive sustainable success, and significantly reduce the risks of major missteps.

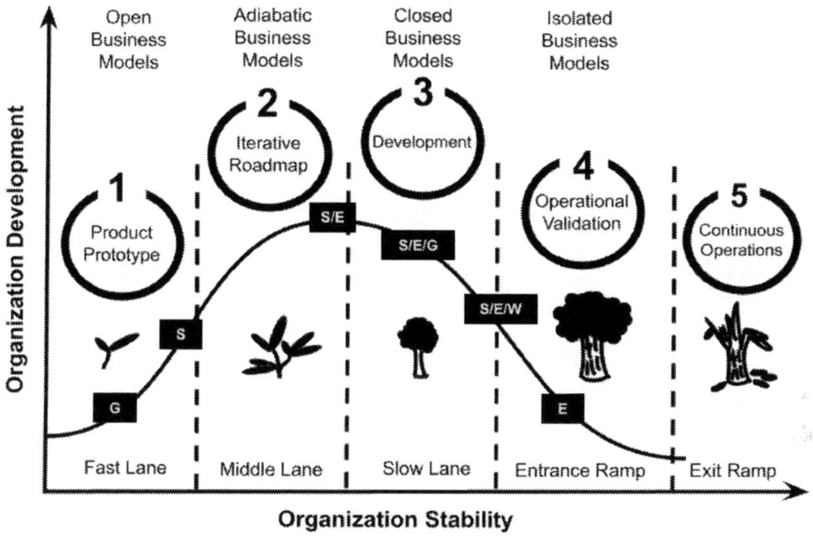

Figure 44. Mechanical System Lifecycle Roadmap to Success

The *Mechanical Systems Lifecycle* is fundamentally about alignment, with a few essential prerequisites to set your organization up for success. First, ensure your company's vision and values align, creating a unified purpose. Second, position your organization to secure the necessary resources—staff, technology, and capital—ensuring that these assets are readily available to support growth. Third, align the organization to cultivate core capabilities crucial for executing your strategy effectively. Once these three foundational elements are in place, your organization is primed to implement the *MES*, setting the stage for sustained success.

Not all opportunities and capabilities are created equally.

The First Principle of Thermodynamics

The *Mechanical Systems Lifecycle* is rooted in the core principles of thermodynamics, prioritizing energy conservation, order of operation, and high quality. The highest-performing business models, products, markets, and strategies remain inaccessible to the masses—often because they require discerning quality, investing time, or paying a premium. I have repurposed **Figure 44** to offer a different lens in **Figure 45**.

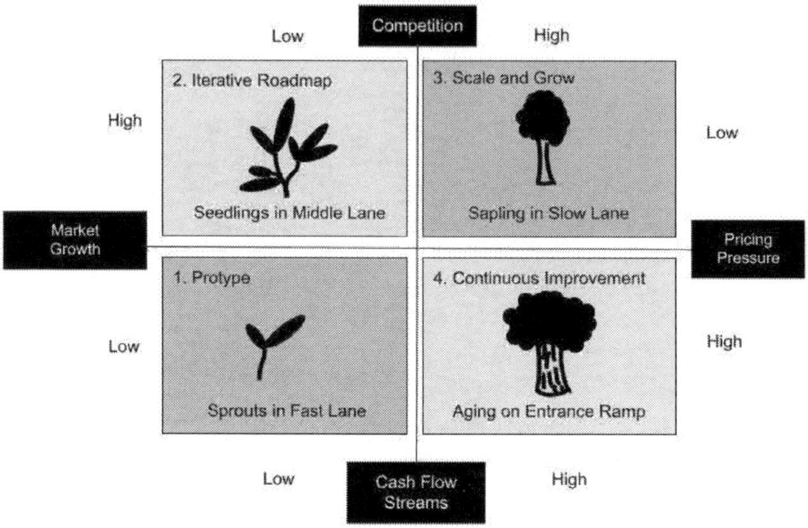

Figure 45. Lifecycle Opportunity vs Capabilities

Many people settle for "good enough," crowding into the average. Real value, however, demands an investment. Yet this up-front investment saves time and effort in the long run and translates into impact and progress in your business. The investment benefits include efficiency, values, and lasting results, all of which are universally desirable outcomes, but 98% of leaders are unwilling to pay the price that high performance requires to rise beyond being average. This mindset is what separates the exceptional from the ordinary. The road to success is steep. Reaching the top is far less crowded and offers more space, less competition, and increased opportunities. The top is reserved for the 2% of leaders who accept the challenge, pay the price of admission, and do what's required to obtain a high-performance organizational system.

Over the next four chapters, we will take a closer look at your Mechanical Engine System to learn more about the four strategy lifecycles: business models, business products, business markets, and business execution lifecycles.

40. Jahirul, Mohammed I. et al. (2013). *The Use of Artificial Neural Networks for Identifying Sustainable Biodiesel Feedstocks* on ResearchGate. Available from: https://www.researchgate.net/figure/Schematic-diagram-of-a-typical-diesel-engine-fuel-system-12_fig6_277503095 [accessed 17 Sep 2024].
41. How Trees Grow. Texas A&M. http://texastreeid.tamu.edu/content/howTreesGrow/
42. Organization Development-Stability Lifecycle Adapted from Xiaojie Li.
43. The Use of Artificial Neural Networks for Identifying Sustainable Biodiesel Feedstocks

CHAPTER 20: Business Models

Some people think that building a business that makes a lot of money is hard. But the real challenge often lies in generating steady revenue over an extended period. Whether you view business growth or restructuring as easy or difficult, you'll never reach your goals or uncover the truth without the selecting the right fuel for your chosen business vehicle. In the *Mechanical Fuel System (MFS),* fuel powers the Mechanical Engine System. Here, the analogy of Fuel (see **Figure 46**) represents the business model lifecycle strategy. The goal is to select the right business model that aligns your business to the right opportunities and capabilities.

Fuel Represents the Business Model

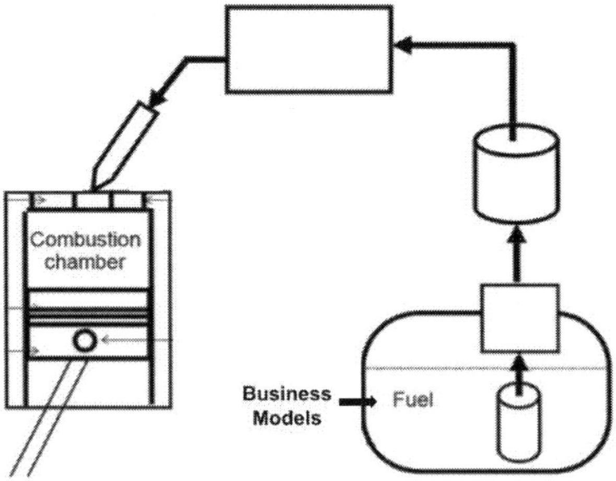

Figure 46. Mechanical Fuel Systems: Business Models is the Fuel

There are different types of fuels, just as there are different types of business models. Not all business models are created equally. This is why following a strategy to help you choose the right fuel is about understanding the requirements for a strong business model.

I'll share the science that supports the four key business models we're discussing, introduce you to the business model lifecycle, and break down the requirements that differentiate each business model. While all have different effects, one business model drains more energy than it's worth. I recommend that most leaders stay away from it. But the other three are worth analyzing. Of these three, only one stands above the rest, creating longevity, legacy, and lifestyle that aligns with becoming a happy high achiever and vital leader.

Your first step in executing the Mechanical Master Plan strategy is to decide which business model will help you implement your business vision. Get this wrong, and you'll struggle. Choose wisely, and you'll benefit from a powerful, efficient business model that fuels your business needs and allows you to thrive.

Business Model System Types Based on Science

The science that supports selecting the right business model is Thermodynamics. The principles of thermodynamics are the laws of energy conservation and order. The law of conservation states energy is finite, which you can only get from a high-source environment. The law of order states that energy is always slipping into disorder. We cannot stop disorder, but we can manage it and lessen its effects on our systems.

Thermodynamic systems come in two types: open and closed systems. These systems can be further categorized into four types: isolated, adiabatic, closed, and open systems (see **Figure 47**). Now, let's define each system.

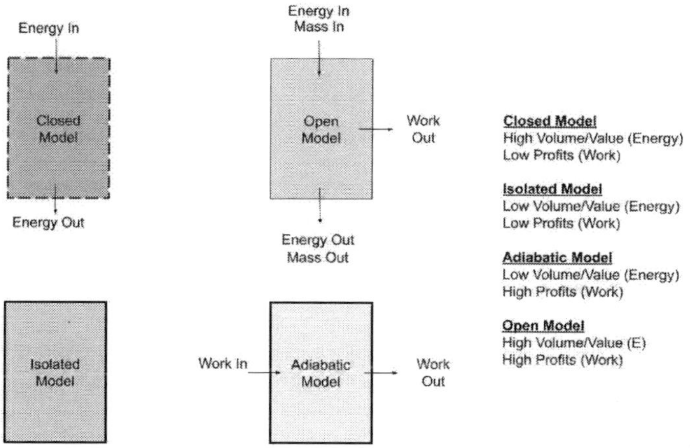

Figure 47. The Science Behind the Four Business Models

1. **Isolated Systems** – These systems are completely insulated, with no exchange of energy or materials with the external environment. It represents an aging lifecycle.
2. **Closed Systems** – These systems have a fixed material content, but they can undergo internal changes, like chemical reactions, that impact their composition.
3. **Adiabatic Systems** – This type of system allows work interactions only, without any heat exchange with the surroundings, maintaining a controlled internal environment.
4. **Open Systems** – In an open system, there's the free exchange of energy and materials with the surroundings, allowing adaptation and responsiveness to external changes.

Now, let's adapt system thermodynamics to business models.

Business Model Lifecycle

When selecting the right business model, think of it like selecting the right fuel for your vehicle. To start, you may want to check the owner's manual to confirm if the fuel system needs a premium fuel blend for optimal performance or if standard fuel will do. High-performance engines require high-octane fuels. The same principle applies to your organization. Remember, you're building a premium business that needs a customized approach to maintain peak results. We've already established that performance is your desired output over the required input, so choosing the correct business model is essential to "fuel" your success.

In **Figure 48**, the lifecycle development is aligned with the business model.

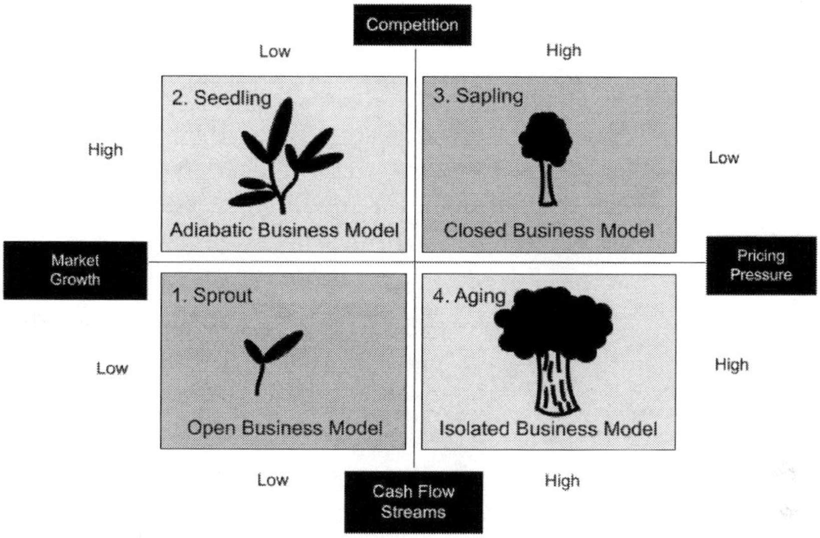

Figure 48. Business Model Lifecycle

Sprout represents Open Business Models, seedling represents Adiabatic Business Models, sapling represents Closed Business Models, and aging represents Isolated Business Models. Before I explain each lifecycle phase, let's look at the science that supports the requirements for selecting the right business model.

Business Model Selection

Before learning about the business model strategy, it's important to understand that not all business models are created equally. The strategy requires selecting the best business model based on its opportunities and capabilities. Business opportunities range from increasing to decreasing pricing pressures, market growth, cash flow streams of revenue, and competition. Business capabilities are the differentiable traits ranging from generic to unique.

Isolated Business Model: Low Volume (LV), Low Profit (LP)

Isolated Business Models are often low-volume and low-profit setups, marked as LV x LP. This model represents the commodities zone.

The Isolated Business Model attracts generalists or jacks-of-all-trades who are eager to begin but find themselves isolated, operating with minimal resources. This model starts as a "test zone." The goal for the entrepreneur is to discover what works while keeping costs low, so they focus on low-margin products. The problem is that many people who start their businesses here get stuck, never moving forward to turn a profit because the margins are too low to lift the business. The result is often a drain of time, energy, and resources. You're a hobbyist, not an entrepreneur, until you start providing value and generating profits.

Isolated Business Models violate the first law of thermodynamics and attempt to extract energy from low-energy sources, often through cheap labor and minimal investments. Sales are slow, so they never build enough momentum to test ideas or divert resources to build capabilities and find new opportunities. This model runs inefficiently, adhering to old-school methods and relying heavily on guarding implementation forces. This business model is often used by startup solo entrepreneurs or small family businesses. Without high-quality energy or robust resources, these businesses lack the energy needed for growth and innovation.

In isolation, there's no transfer of energy (value) or mass (growth). Time becomes a constraint, as resources and assets are quickly lost due to not having structure, processes, or systems. Isolated businesses compete on low margins and volume, vying for market scraps. Strategies are often based on low-value, commodity-based thinking and shy away from competitors. Consequently, they lack fresh ideas and market understanding, and their distribution channels fail to support consistent growth. What happens is that Isolated Business Models go up against well-entrenched vendors who have a complete service offering, so you get crushed.

Isolated Systems Create an Error Before You Start

An isolated, self-contained business creates an error before you start. This business model attracts leaders who focus on a problem no one cares about. Most valuable payers (MVPs) are unaware that they should care about it, they don't value the solution, or the problem has already been solved with an alternative solution. As shown in **Figure 49**, the business starts with an idea or prototype but goes straight into operating within the commodity zone of an aging business model. It never established margins, never built awareness, and never established thought leadership needed to differentiate itself. I do not recommend this business model to my clients starting a new business or seeking to improve organizational performance. But if you find yourself here, **theoretically**, you can prototype a product into a commodity market. You must first show thought leadership for why your approach is different, then take the time to follow the iterative roadmap for Middle Lane customers, then go into the scale for growth, and not from prototyping directly into a commodity market.

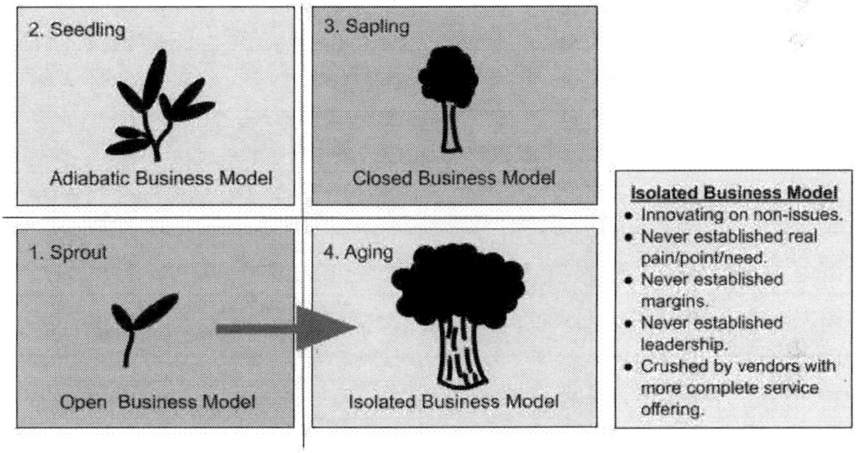

Figure 49. Isolated Business Model

Closed Business Model: High Volume (HV), Low Profit (LP)

Closed business models are often regarded as high-volume, low-profit setups, represented as HV x LP for high volume and low profit. These models focus on scaling for quantity rather than quality, which enables making a profit through volume. Yet, this model is saturated and heavily favored by large enterprises, making it challenging for smaller businesses to compete. Think of it as a scaled-up version of isolated models, with heightened pressure on profit margins and limited space for innovation—a business setup with this model often falls short of long-term success. Closed business models attract entrepreneurs who want to compete on the big stage or obtain moderate business success in previous ventures and have the capital to pay to play.

Despite low-profit margins, closed models can generate massive revenue through careful margin management. Larger companies pass expenses on to their suppliers, subcontractors, or end customers to offset operational costs. This "too big to fail" mentality often limits creativity, focusing instead on cost reduction, shrinkflation, or price increases to maintain profitability. Over time, this cost-focused mindset curbs innovation, stunting growth and industrial evolution.

Companies that thrive with a closed model leverage vast supply chains and distribution networks, positioning them in a league of their own. Giants like Walmart and Amazon exemplify this approach, with their extensive infrastructures enabling them to deliver high-demand products and services. These organizations are mature. They tend to collaborate extensively, subcontract tasks, and rely on their partners and suppliers to help them meet the high-volume demands of this model. While these companies create significant value through volume, they also demand immense resources—technology, skilled labor, research, assets, energy, time, and warehouse space. Their competitive edge is often their ability to scale a micro-market they carved out within the larger ecosystem, strengthening their market position.

Closed System Loses Oxygen and Flames Out

I do not recommend Closed Business Models to my clients starting a new business or seeking to improve organizational performance because these businesses eventually lose oxygen and flame out (see **Figure 50)**.

What causes a closed business model to lose oxygen and flame out? What happens here is a very visionary entrepreneur tells a compelling story and secures a lot of financing or secured financing from selling their last company. Their vision is so clear that they build every feature and function under the sun. They have the "if I build it, they will come" syndrome.

They go for a global release right out of the gate. They scale it too early and don't take the time to innovate or understand the core customer's pain or journey. They fail to understand the spending priorities of their MVPs, and they don't have the documented evidence to support their claims and decisions. So, these business leaders assume demand exists, do a big launch with lots of noise, build lots of excitement, and, poof! They flop. They've built a business based only on their ideas and assumptions—a closed system with no oxygen. The flame of success doesn't hold up over time, and it flames out.

One downfall of closed business systems is that they fail to build documented evidence proving the core customer problem has been solved. Instead, they grow and scale up too soon and compete for opportunities without building capabilities.

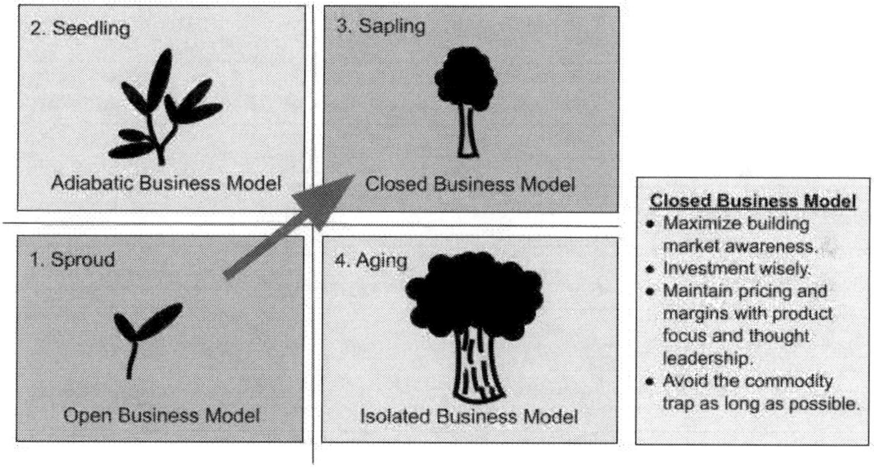

Figure 50. Closed Business Model Lifecycle

Adiabatic Business Model: Low Volume (LV), High Profit (HP)

Adiabatic Business Models represent a unique, high-profit approach, marked as LV x HP for low value, high profit. This model is where innovative products and services can gain traction, though they often require a precise strategy to maintain steady growth.

In an Adiabatic Business Model, while energy is transferred

through work, there's no actual mass transfer, creating conditions ripe for sporadic success but lacking long-term stability. Dr. Boyd F. Edwards' Adiabatic Heating Demo Video on YouTube serves as a helpful illustration of this process, showing how an adiabatic approach generates energy without lasting fuel transfer. This model can be an excellent starting point, but it quickly needs refinement to create sustainable growth.

The adiabatic phase is often where entrepreneurship finds its first spark. High-ticket sales can appear promising, but without a steady flow or clear market positioning, growth becomes challenging. Your offering may have immense potential, but to unlock its full value, you must refine your strategy, strengthen brand clarity, build through leadership, and improve product discoverability. Many promising entrepreneurs stalled here—trapped by inconsistent revenue and unclear positioning. Seeking professional guidance to refine your product and market fit is crucial. With the proper training and education, you can present your value to your most valuable payers (MVPs), expand your market reach, and increase sales.

As a type of isolated system, an adiabatic business treats its product as experimental, requiring more fine-tuning and market testing. If you're selling a high-ticket product but struggling to make sales and scale, it's time to reconsider your business approach. You can strategically re-position your offerings and prove your value proposition with a few tweaks.

Adiabatic Business Models rely on seed capital from personal assets or initial investments from family and friends who trust you. Friends and family support your venture because they believe in your capabilities, not necessarily in your business concept. This foundational trust provides the seed capital needed to start. It is also possible for angel investors to infuse additional capital into the business to help maintain internal business operations during dry spells.

As I mentioned earlier, most entrepreneurs start their businesses here. If you're in an adiabatic phase, consider this the foundation of something greater. With the right adjustments, you can transform this flickering flame into a steady glow, propelling your business toward the stability and growth it needs to thrive.

Adiabatic System is a Lost Opportunity

So, what happens when an entrepreneur starts right but loses their opportunity along the way?

Well, let's start by covering what they did right, as illustrated in **Figure 51**. They started by prototyping their offering to their most valuable payers (MVPs) and charged a premium. They have a good core product offering but don't have enough oxygen to sustain the business. Their marketing efforts are inconsistent, or their messaging falls short. They fail to build awareness, and the long, dry spells between sales slow growth. Therefore, they can't execute quickly. The business is like a shooting star.

What causes this? Well, maybe there's a lot of entropy and disorder in the business—fighting between teams or co-founders. Maybe they just can't capture the opportunity because there's too much inertia or friction in the business, and they lose the business opportunity. As a result of the internal issues, they fail to establish thought leadership and documented evidence. Basically, they can't get out of their own way and ask for help. By following the prescribed strategies in Mindset Mechanics, they have a fighting chance to save the business. What's sad is that adiabatic business issues happen often, and many entrepreneurs and leaders find out too late the cause and that a proven solution exists.

When this occurs in businesses, you'll see that rather than just killing the business, trying again, or pivoting, entrepreneurs end up dropping their prices and being low-cost providers out of desperation. They place a great product into the commodity zone. They think, "Some revenue is better than no revenue," and give away their value. As a result of taking this action, their passion, vision, and innovation die with the business. Once the heart and character of the organization fade, it's like dying fruit on the vine. It's a lost opportunity.

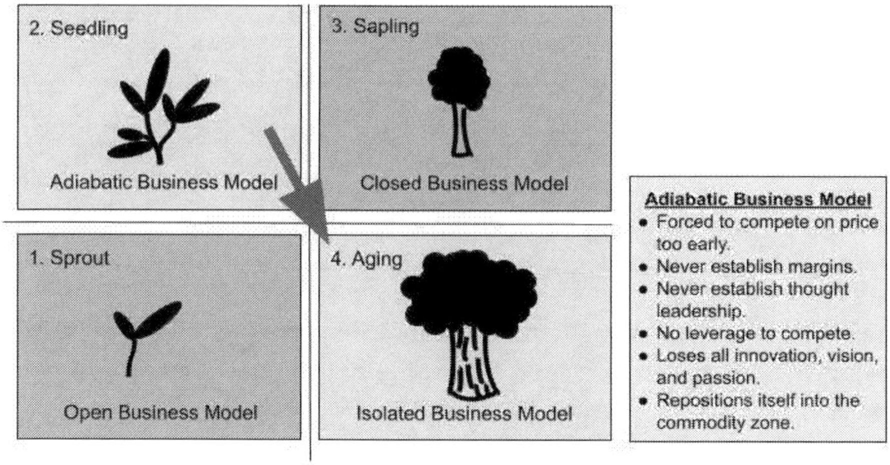

Figure 51. Adiabatic Business Model Lifecycle

Open Business Model: High Value (HV), High Profit (LP)

This final business model is how I've built my consulting practice and from where I teach my clients. Open business models represent the peak of business strategy, combining high value (HV) and high profit (HP).

Adopting an Open Business Model is more than a wise choice—it's the game-changer that positions your business for elevated success. Open Business Models provide the perfect platform for launching products and services primed for success and positioned for performance. Open Business Models excel because they're designed for true differentiation. By understanding markets, customer needs, and competition, businesses can integrate seamlessly, foster innovation, and deliver exceptional quality. Moreover, these models leverage strategic partnerships, collaborating with other top-tier businesses to elevate the marketplace as a whole. This collaborative approach builds a thriving ecosystem, and in the upcoming section, we'll explore what makes this possible with the Business Model Matrix.

Open Business Models attract and retain top talent, ensuring smooth operations and setting the standard other companies admire. This business model attracts the 2% of vital leaders who are happy high achievers. They want to build a fulfilling, purpose-driven lifestyle by leveraging premium resources and driving exceptional performance. Selling within this model is far more advantageous than the other three business model approaches. This business model is for leaders who prioritize differentiating capabilities and harnessing new opportunities.

If you think excelling in an Open Business Model is difficult, it's time to reconsider. It takes no more effort than any of the other business modes, but it does take thinking differently. Delivering valuable results is essential, and once you understand the mechanics, you'll wonder why you didn't start here sooner. This is your moment to seize the opportunity and set your business on the path to remarkable success.

Open Systems Take the Long Way Around

The Open Business Model is a Roadmap for Success designed to maximize your market awareness, make smart, timely investments in your time, and strategically manage your growth. It avoids all the other pitfalls we discussed. The Open Business Model requires that you build thought leadership early for your most valuable prospects (MVPs). You must also develop and refine offerings to demonstrate an understanding of your MVP journey, pain, and spending. When the time is right, you grow and scale to sustain results in competitive markets and avoid stagnation by carefully managing your aging offering (see **Figure 52**).

For Open Business Models, go around the curve. This method ensures you maintain the strategy, pricing, and margins throughout the lifecycle. The curved path focuses on the product and customer and builds name recognition, influence, and authority at every stage. Open Business Models, when followed and executed seamlessly, are designed to avoid falling into the commodity trap, where your business becomes indistinguishable from your competition and overly reliant on price competition. By executing the roadmap correctly, you create a resilient business capable of adapting to every lifecycle stage while driving long-term profitability and growth.

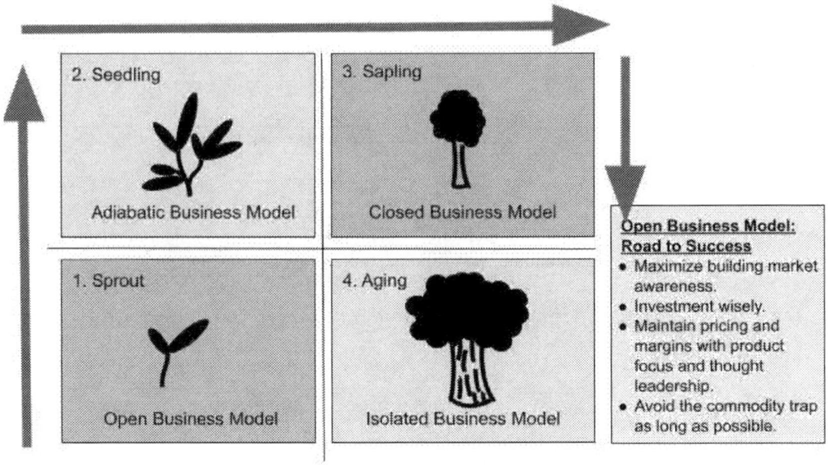

Figure 52. Open Business Model Lifecycle

Summary Of Business Models

The Profit-Value Differentiation Matrix below allows us to view each business model from a different lens to compare and contrast their value. As you can see, it's the open business model that provides you with the best opportunity for success.

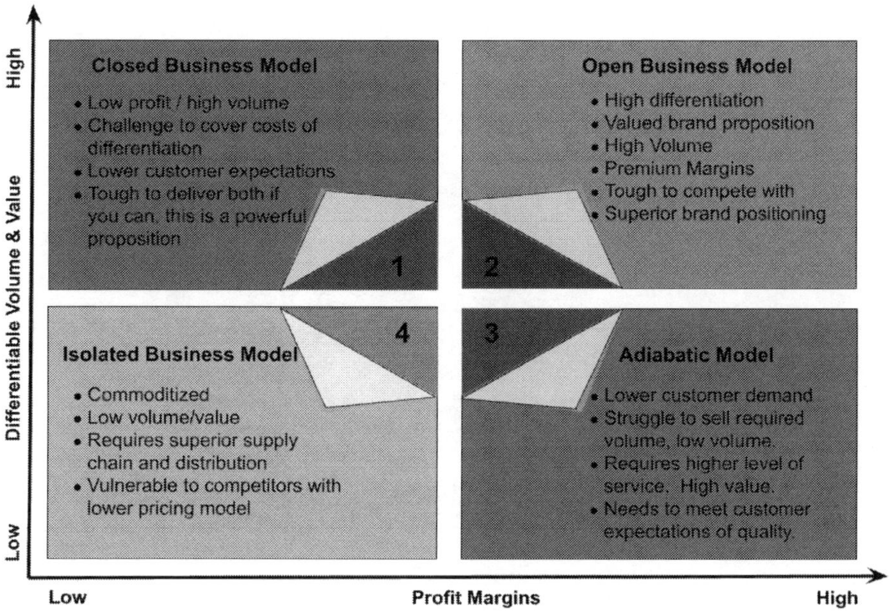

Figure 53. Profit-Value Differentiation Matrix

Business Model Financing

As you're going a long way around the curve, it can be helpful sometimes to conceptualize where your sources of financing will come from. Let's look at **Figure 54**. As we discuss it, remember that when your business unit is on the left side of this chart, you have fewer cash flow streams of revenue or financing available. As you progress through the financing model, access to capital increases; I recommend accessing the best match at key times throughout the business model lifecycle.

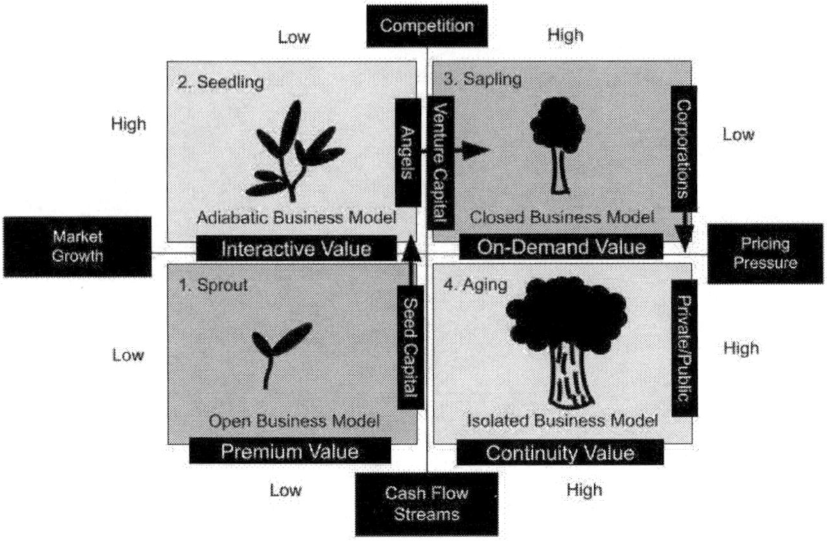

Figure 54. Business Model Financing

Seed Capital

Business financing starts in the sprouting phase of business development with internal funding sources, MVPs, and professional seed capitalists. Initial investment takes on the form of time or funding from personal savings, friends, or family. Trust equity becomes a valuable currency for seed capital investors.

Friends and family support your venture because they believe in your capabilities, not necessarily in your business concept's proven success. Foundational trust provides essential financial support, though it's imperative to convert this into a lasting market presence.

Here, seed capitalist motivations shift from family and friends who trust you to MVPS, who trust your business concept and ability to create innovation. The revenues in the sprout phase of business development are generated from premium sales. Premium sales help sustain the business's internal needs, allowing you to focus on understanding the MVP's pain, journey, and spending limits. It also allows you to build documented evidence, thought leadership, and trust.

Professional seed capitalists invest in you because they believe in the vision. They know where they are going. They can clearly see what direction you're headed in, and they want to share the journey. Whether seed capital is obtained from savings, family and friends, MVPs, or professionals, seed capital is your best source of startup financing.

Angel Investors

As you continue to grow as a thought leader, it's time to start focusing on investors who specialize in funding companies in the seedling phase of development. Angel investors are your seedling MVPs or professionals. Angel investors in this phase don't fully understand the big picture, but they will also trust and invest in you if given a reason. They don't have everything figured out because some roadblocks block their understanding and view, but they're willing to learn and make a bet that your business concept will solve their immediate problems. Angel investors provide capital that will help you iterate your prototype and really get the documented evidence you need during research and development.

Venture Capitalist

Angel investors make introductions and share your business story or their experience with your business with Venture Capitalists (VCs). VCs specialize in sharing products with evidence that your business concept understands the customer's journey. To attract VCs, your business needs to have the right team, introductions, structure, and liquid cash flow to start to really aggressively move into growth scale mode to become a maturing business with solid marketing activities.

Corporate Investors

Once you've hit early-growth scale mode, you'll typically get the interest of corporate investors who want to fill out their own product market suite of offerings, sponsor deals, partner with consultants, or license your product or service. Sometimes, corporate investors will invest early while you're still in the seedling development phase. When this happens, it's really an R&D purchase. But if you can get the messaging right and put fear corporations that if they don't partner with you, you'll eat their lunch with your fully baked solution, offering, and getting lots of market demand. They'll try to buy you out or take you out to eliminate a future competitor and add to their own product market solutions.

Private Equity / Public Partnerships

Then, there are private equity and public partnerships. Private equity and public partnerships typically play down in the late entrance ramp into the commodity zone. Here, it's all about discounts and financial factoring.

If you wait for public partnerships to contact you and you can show differentiable value, you can increase your value and be designated as a sole source provider. Things like that now aren't always true, but this is a general representation. If you can show differentiation and show up during a crisis, you can increase your perceived value and be designated as a sole source provider.

For example, financiers compete in a market with scarce opportunity and finite energy, so if they're having a hard time with their core offering, you might see a private equity firm acting more like a VC. A VC acting more like an angel. You might even see a government agency acting more like a corporation. Typically, the Business Model Financing in **Figure 54** shows what type of financing you should pursue at different stages of the business model lifecycle.

Market Value and Ecosystem Creation

In today's evolving business landscape, fostering a robust business ecosystem is no longer optional—it's essential. This approach goes far beyond traditional partnerships. It involves transforming your organization into a dynamic marketplace that actively serves clients, teams, and partners' needs. Embracing a business ecosystem enables enhanced collaboration and drives substantial value for all participants, strengthening the marketplace as a whole.

A compelling example of this approach comes from Kati Brock,

PhD, and her team's comprehensive case study, *Light the Way for Smart Cities: Lessons from Philips Lighting*. Over a five-year period, they analyzed and compared four smart cities, offering invaluable insights into the journey toward smarter, more sustainable urban living, which was adapted into the Business Model Matrix shown in **Figure 55**. Research suggests that open businesses are the crucial players in building sustainability for smart cities. The research also revealed traditional, product-centric business models must evolve to keep pace with innovative, transformative advancements.

By embracing the Business Model Matrix, innovative businesses can unlock the full commercial potential of their advancements. This shift facilitates technological progress and fuels significant corporate transformations, positioning companies for revitalized growth and long-term impact in a rapidly changing market.

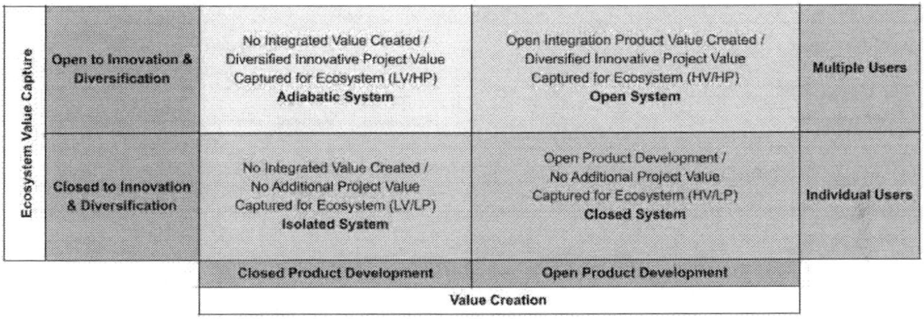

Figure 55. Business Model Matrix. Contrasts 4 Distinct Types of Marketplace Value Creation and Capture

Open Business Models set themselves apart by capturing ecosystem value, welcoming innovation, embracing diversification and differentiation, establishing thought leadership, focusing on user-centered development, and seeking new opportunities. Closed business models restrict innovation and limit collaboration, but Open Business Models position organizations to thrive by engaging with a broader range of stakeholders. For leaders, understanding that sticking to outdated business models can stifle adaptation and slow the creation of valuable products is critical. This resistance, often driven by an attachment to traditional practices, risks missed opportunities in today's rapidly changing markets. Leaders who adopt an ecosystem mindset, recognizing the interdependence of upstream and downstream organizations, can find an effective balance between cooperation and competition that enhances their strategic advantage.

The distinction between these models lies in their focal point. Traditional models focus solely on direct value exchanges between the business and most valuable payers (MVPs). However, an ecosystem approach broadens the lens, incorporating essential value exchanges outside of the MVPs. This broader view is crucial for delivering sustainable solutions, much like those seen in the smart city movement. By embracing an ecosystem perspective, leaders can drive innovation and effectively meet community needs. This creates an ecosystem outside the macroeconomy, protected from market volatility, and created to sustain a subculture of market diversity and constant innovation.

The smart city case study offers valuable lessons for ecosystem-building, demonstrating how interconnected value exchanges create sustainable growth. Examining the Business Model Matrix reveals four distinct business model types, which can serve as a foundation for developing and capturing marketplace value.

Picture your business model as a flourishing smart city—a dynamic ecosystem of minimum viable products that bolster local economies. Vital leaders understand the importance of fostering an environment that enables the organization and ecosystem to prosper.

---

44. Dincer, Ibrahim, Four Business models developed from *Thermodynamic Systems: Fundamentals*. Geothermal Energy Systems, 2021.
45. Demirel, Yaşar, Fundamentals of Equilibrium Thermodynamics. *Nonequilibrium Thermodynamics*. 4th Edition. 2019. https://www.sciencedirect.com/topics/engineering/thermodynamic-system
46. How the Product Development Process Works in 5 Steps https://www.scalefocus.com/blog/how-the-product-development-process-works-in-5-steps

CHAPTER 21: Business Products

In business, just like in a high-performance vehicle, the fuel system is more than a simple component. It's an intricate system that keeps everything moving forward. In the *Mechanical Fuel System (MFS)* framework, the fuel system doesn't begin at the engine—it starts at the tank. Though the process is straightforward in a car, for us, this represents a reservoir of assets and offerings that fuel organizational growth. The fuel tank and pump work together to channel energy, make products and services available to the marketplace, and turn ideas into tangible success (see **Figure 56**).

If you want to get technical, fueling begins when you remove the gas cap and fill the tank. But for our framework, the tank serves a broader role. The fuel tank stores all the products and assets for your business. Once this fuel is picked up by the fuel pump, the real journey begins. In our model, the fuel pump represents your products and services, ensuring they are ready and available for customers whenever needed.

Fuel Tank and Fuel Pump Represents Business Products

Fuel pumps must operate consistently to ensure fuel is supplied. Just as engines rely on pumps to keep them running efficiently, your strategy depends on pumping products into the market. In the MFS, we determined in Chapter 20 that selecting the right business model is your business fuel, while the fuel tank and fuel pump represent your suite of products and services. Stored within the "fuel tank" of your business, these products and services are drawn to meet the demands of your MVPs, create value, and propel your organization forward.

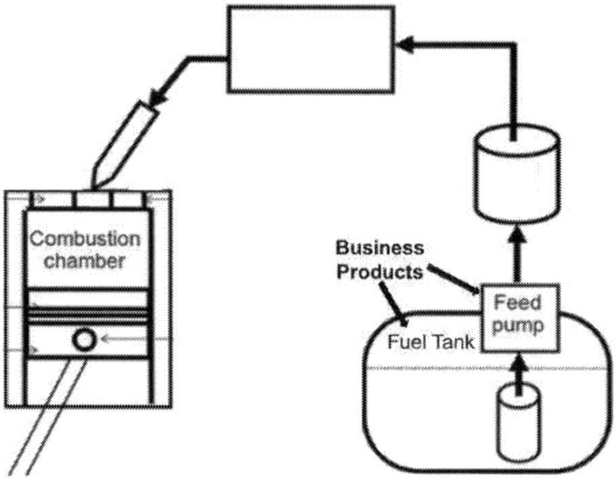

Figure 56. Mechanical Fuel System: Fuel Tank and Pump are Business Products

When the fuel pump operates effectively, it consistently draws from this tank of offerings, ensuring that your products are always accessible and positioned to meet market needs. Whether your organization operates with a single product or a diverse suite, the key is keeping the pump functioning optimally so products pump seamlessly from the fuel tank into the hands of the customer.

Product Lifecycle

To maintain a thriving and sustainable business, your offering goes through the product lifecycle represented in five stages: the product prototype, iterative roadmap, development, operational validation, and continuous operations. Passage through these stages forms the backbone of a reliable product. It ensures that your offering has met the needs of an evolving environment. Executing each phase of the product lifecycle keeps your business moving forward (see **Figure 57**).

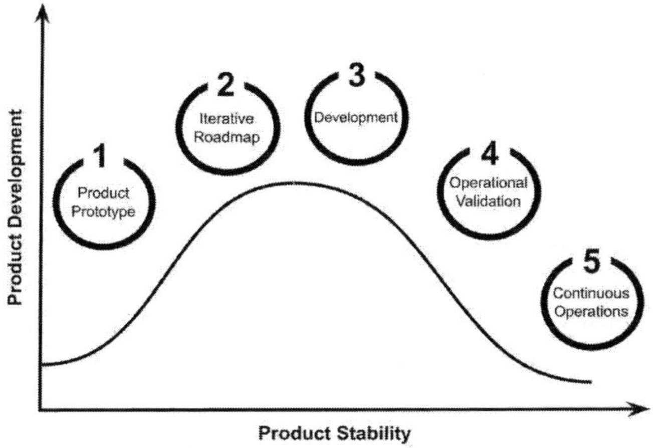

Figure 57. Product Lifecycle Stages

Product Prototype Phase

In the Product Prototype Phase, you refine and polish your product strategy. This is where preparation meets precision. Imagine yourself as a dedicated scientist, questioning every assumption and challenging established norms. What's the goal of this phase? To establish authority, research influencing products in the market, and understand the power dynamics of a competitive landscape. To achieve this, gathering strategic information serves two essential purposes. First, this allows you to position yourself as a thought leader by sharing breakthroughs, successes, and solutions, showcasing your innovative approach as superior, and clearly communicating why it excels. The second purpose is to craft a compelling vision for the future that attracts high-caliber clients, partners, and vendors—people eager for insights and solutions that align with cutting-edge market trends.

In this phase, your objective is to draw in your most valuable prospects (MVPs)—the proactive, forward-thinkers who not only resonate with your vision but will actively champion it because it reflects their own values. These are your ideal premium clients, seed investors, and supporters who expand and fuel your open business model.

The product prototype phase is the moment to *spout* innovation. If you do this right, your offer should challenge and disrupt the status quo. Successful innovation creates waves in the marketplace, capturing the attention and interests of your target audience actively looking for innovation. Once you get their attention, dive deep into painting the picture of how your solution challenges the dynamics for your niche. This means you've identified an existing challenge, discovered unanswered questions and unmet needs, and have access to the prevailing solution. Where are the gaps in the market? How are MVPs approaching these issues? How does the MVP describe the pain and see the issues? Understanding your MVPs' mindset, language patterns, and journey is essential to delivering a premium product that not only meets their needs but anticipates them, setting you on a definitive path for iterating product value.

Iterative Roadmap Phase

The Iterative Roadmap Phase focuses on interacting with MVPs in the *seedling* phase of development. Shifting into iterative thinking plays a key role in this phase. You must work to truly understand the pain and customer journey, build influence, and document the results that prove you've captured value and created a solution. MVPs on the iterative roadmap value productization, the packaging of products and services for consumption. To achieve productization, start by proving your premium offering is tailored to specifically address core problems and engage, educate, and alleviate challenges without requiring the full investment of a premium solution. Offering productization offers them a path for growth, allowing solopreneurs and small businesses to access what they need when needed. This requires breaking down your core product into digestible products and services. For this phase, imagine yourself as an investigative journalist, digging deep to discover the "why" and identify the real problems your customers face so they can be addressed individually. By isolating the issue, it's easier to figure out what works so you can continue to refine it and deliver the best value. Adding value is accomplished by incremental "adiabatic sparks" of improvement to your product, keeping it lean and providing constant feedback over time.

With each iteration, you measure the outcomes because growth is tied to how well you document and understand your customers. As they say, until it's documented, it's never done. Deliver consistent insights, cultivate relationships, build credibility, and establish a record of trust needed to carry you through this phase. MVP customers in this phase may not be ready for your premium offer because they have immediate needs that, once addressed, will cause them to return for additional productized offerings and be open to exploring broader solutions.

Working closely with your MVPs is essential to ensuring your solution genuinely meets market needs. Collaborate to validate that your solution addresses their core issues and document the outcomes. The primary aim is to build strong, enduring relationships by demonstrating your commitment to giving their pain a voice and solving their problems. Success in this phase will be marked by repeat purchases, contract renewals, and growth with these MVPs—proof that you've crafted a roadmap to success they can rely on.

Development Phase

The Development Phase is about maturation and establishing your brand as a recognized, trusted leader. This is the *sapling* phase. Here, your target MVP will only consider purchasing from a trusted authority. In the development phase, you are a skillful director overseeing the various points of interest and creating products and services that keep the MVP on track for positive growth and keep them focused on the right things.

By now, you've been building up thought leadership equity. It's time to leverage this equity of proven product results and reputation to establish a commanding presence as a sapling tree in your niche or industry. Trigger this phase prematurely, and the existing competition will quickly pounce and expose your business as unprepared to compete against giants. In this phase, you may start seeking partnerships with large networks, authoritative brands, and industry leaders to reinforce your credibility and amplify reach, but be sure you have something of value to offer them. The focus here is on delivering high-quality and expanded distribution capabilities to grow and scale up. The development phase measures how well you can effectively boost volume to meet demand.

The challenge in this stage is twofold and ripe with opportunity. First, develop a universal, stable, and standardized product that caters to the mass market. It should allow you to capitalize on economies of scale, increase margin, decrease production costs, and boost operational efficiency. Secondly, supplementary products and services should be created to address the specific needs of an evolving yet stable MVP, meeting them where they are on their journey.

This is the ideal time to build or integrate a comprehensive ecosystem of products or services that complement and evolve with your core offerings. You should also consider collaborating with external suppliers, strategic partners, and vendors that can add further value, helping you broaden your product line or explore new applications for existing solutions. The ultimate aim here is to maximize profit margins, significantly increase volume, and aggressively capture market share among more established MVPs to position your brand as the go-to leader in your industry.

Operational Validation Phase

In this phase, operational validation ensures that your product's output meets the intended purpose with reliable accuracy. The goal is to maintain product appeal for the next tier of MVPs—those clients are transitioning into the entrance ramp as an aging demographic buyer. Here, you can consider yourself a skilled groundskeeper dedicated to preserving and enhancing your established market position by capitalizing on your expertise, proven systems, and industry standards.

At this stage, the product reaches the *aging* phase. Your focus is not on disrupting the markets but on careful management and consistent operations. Nourish the product and your existing clients, and avoid additional investments or innovations during this phase. Prioritize investments that promise immediate returns; if a new development lacks an immediate profit path, consider postponing it. Continuously evaluate the returns and operational validation against the possibility of winding down the product line, weighing market needs carefully.

This phase's primary objective is attracting cautious new clients on the market's entry ramp that value low risk. You will leverage your cash flow, brand authority, and established resources to sustain growth. Your product may be aging, but emphasizing its proven value, track record of success, and reliability allows you to appeal to a new segment of MVPs looking for dependable solutions. Leverage case studies, testimonials, and trending industry data to showcase the product's strength and relevance, solidifying its appeal as an indispensable choice for prospective clients.

It's important to note that MVPs in the operational validation phase of product development seek low-risk and low-cost solutions. If you've done the other three phases correctly, you operationally validated repackaged and repurposed products and services that deliver value at the lowest costs without wasting excess energy and resources or negatively affecting your bottom line. This also means you've done a great job developing differentiating value and capturing the right opportunities so you don't have to compete based on price. You can meet their needs and maintain profitability, even in commodity-based markets.

Continuous Operations

Continuous Operations is the exit ramp, the final product lifecycle phase. Here, the product is in a transitional, *dying* phase. You're providing life support until a decision can be made. The primary goal is to assess the product's viability, considering profit margins, customer feedback, and overall value. At the end of its lifecycle, it's time to decide whether to rejuvenate the existing product, innovate it, or prepare to end the product's lifecycle. This stage marks the activation of your exit strategy, balancing operational maintenance with the option to either sell or phase out the product line.

If selling off the product is the chosen path, you'll need a dedicated team in place to ensure a seamless operational transition and to provide post-transfer support. On the other hand, if discontinuation is the best course of action, the focus shifts toward executing the final steps of your exit strategy effectively. This entails ensuring that assets are acquired, sold, or disposed of; loose ends are tied up. If the product has active customers, you want to ensure they have been notified, all their needs have been met, and resources are reallocated to fuel future growth opportunities. In the dying phase, assets are community-based resources that may become fuel for new products and services or serve historical value.

Once you've followed the product lifecycle, your next focus is understanding business market lifecycles. Let's explore how you should approach the market to ensure the right people see your offerings.

---

47. Kunchev, Krasimir. "Product Development Process in 5 Steps." *Scalefocus*, 12 January 2023, https://www.scalefocus.com/blog/how-the-product-development-process-works-in-5-steps. Accessed 13 October 2024.
48. Adizes, Ichak K. (2023). *Managing Corporate Lifecycles.*

CHAPTER 22: Business Markets

As we delve into the concept of business market lifecycles, we're examining the specific customers you will reach and sell to over time. Understanding your business market lifecycle is crucial for aligning your offerings with evolving customer needs while maintaining relevance in a competitive environment. In this chapter, we are going to look at the market, market segment, and marketing message.

Fuel Lines Represents Business Markets

In the context of the Mechanical Fuel System, fuel for the *Mechanical Engine System* has reached the fuel filter and fuel lines (see **Figure 58**). Just as fuel is filtered and routed through lines that carry it toward the engine, your product message is filtered and channeled to different market segments in order to reach the right customers for maximum impact.

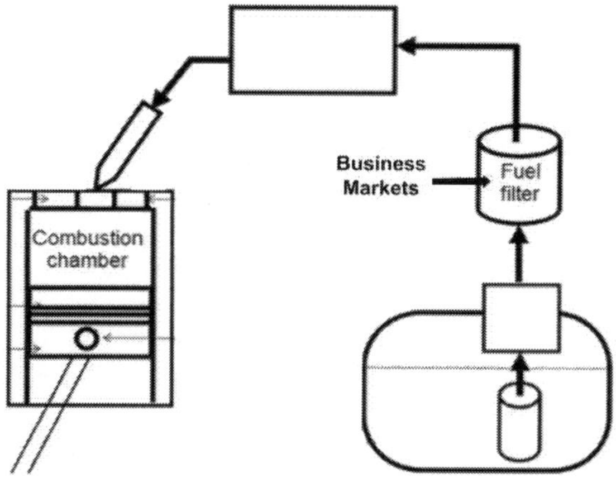

Figure 58. Mechanical Fuel System: Fuel Filter Represents Business Markets

Market segments undergo phases of development and stability, much like business models and product lifecycles, and these markets require different approaches to build momentum. One of the most critical concepts in business marketing is understanding how these markets evolve and communicate. It has challenges, language, and behaviors that can be grouped by development level. Each market segment has a unique personality that is predictable. Understanding market segmentation strategically positions your business by matching the market segment with the right message.

The Business Market Lifecycle provides advantages to understanding segmented market traits to enhance your reach and effectiveness. The business market lifecycle operates separately into four distinct business markets, starting with your core target market and extending to an adjacent outer ring.

Business Market Lifecycle

The Business Market Lifecycle categorizes customers into four types, often referred to as either the most valuable prospects for marketing or the most valuable payers (MVPs) for sales. Just like interstate highways, which have lanes occupied by a host of different driver personality types, business markets consist of different personality types, skills, and maturity levels. Understanding the distinctions between each lane is essential for aligning your product offerings and market messaging (see **Figure 59**).

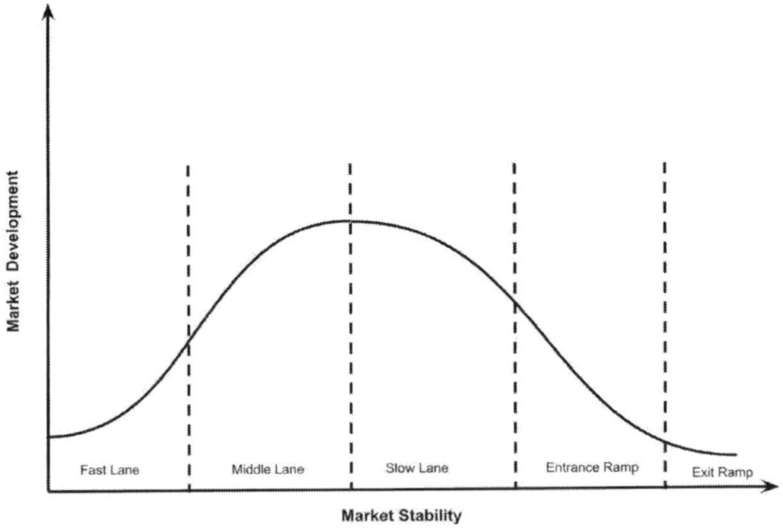

Figure 59. Business Market Lifecycle

Each lane in the Business Market Lifecycle reflects five distinct customer market segment types:

A. **Fast Lane:** The fast lane market segment is eager early adopters seeking innovation and engagement.
B. **Middle Lane:** In the middle lane market segment, you'll find your early adopters seeking a clear roadmap and iterative solutions to address immediate needs.
C. **Slow Lane.** The slow lane market segment prefers to work with established, validated solutions for steady, proven growth.
D. **Entrance Ramp:** The entrance ramp market segment represents the most vulnerable of the four groups, as they only explore market solutions that provide stability and certain reassurance before committing.
E. **Exit Ramp:** The exit ramp represents the end of the market lifecycle. This group is only interested in free, open-source, community-based resources.

To meet each group's unique needs and facilitate their market lifecycle progression, you must tailor your solution, engagement, and personality styles to align with each market segment's problem, leadership expectations, and communication style. Align the customer with the most appropriate message. Strategic alignment fosters growth, builds trust, and maximizes value across the entire Business Market Lifecycle.

Market-to-Development Match

In the business market lifecycle, your organizational message must align with the developmental stage of your product and your market (see **Figure 60**). Just as a young child communicates and behaves in ways appropriate for their age, your organization must similarly adapt its communication and actions to match its stage of development. Then, your offer must be packaged and targeted with the right customer in mind. Moreover, when interacting with other organizations as clients, you must tailor your communication and product offerings to align with the client company's developmental phase, as well.

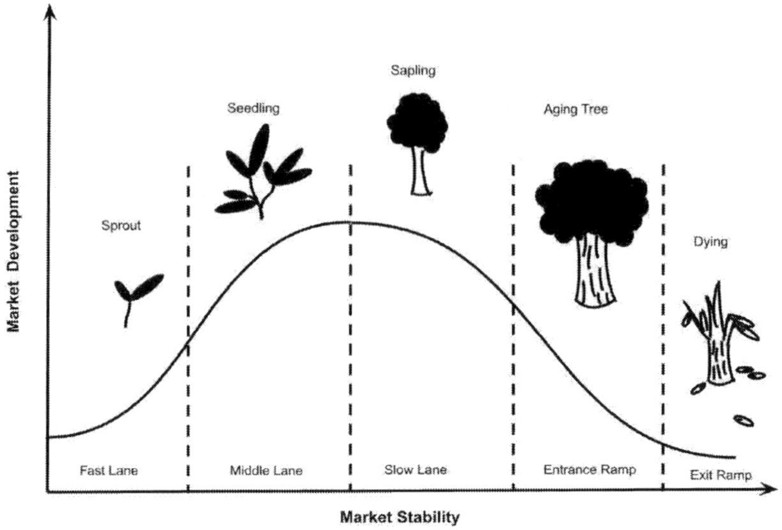

Figure 60. Business Market Lifecycle: Development

During the iterative prototype product lifecycle phase, your messaging should be innovative and future-focused, targeting creators and innovators with the language they resonate with. This is your opportunity to demonstrate thought leadership and establish authoritative command over your product capabilities, market challenges, and industry values. Conversely, in the operational validation product lifecycle phase, showing that you understand the client's business and can nurture the relationship effectively becomes vital. Here, your role will include providing necessary support and assistance and reinforcing trust and collaboration.

Before we delve deeper, let's clarify some key concepts:

Natural Frequency: All systems resonate at a natural frequency, which can be likened to the number of swings per minute. This has fascinating implications for how much influence one can exert on a system. Often, sellers expect customers to buy based on the seller's schedule, but this approach falls flat. Instead, you must identify the buyer's natural buying frequency and align your product or service with their current focus. The natural frequency represented by the bell curve is the target frequency.

MVP's Natural Frequency: Each buyer has a unique natural frequency that dictates their responses. Understanding these frequencies is crucial, as they influence how buyers react. This understanding encompasses recognizing the forces at play and the personality types present within an organization. You need to understand the development of each phase to match the natural frequency.

Message Natural Frequency: Your message also possesses a natural frequency. Merely pushing your message won't enhance its reach; instead, personalizing your message and demonstrating a deep understanding of your MVPs can significantly increase its impact. Tailoring your message to each developmental phase of your product will enhance its amplitude and effectiveness. When the message frequency aligns with the natural frequency, it will be on the bell curve. Otherwise, corrections are needed if the message frequency is under or over the natural frequency.

Markets Natural Frequency: Markets have their own natural frequencies that depend on the growth stage of your client's organization. Recognizing this frequency helps tailor your approach to meet market demands effectively. The market frequency is separated by dotted lines to signify lanes. When you enter a new market, the message must change and adjust with the personality type, message frequency, and natural frequency in mind.

Critical Point of Resonance: When executed correctly, all three frequencies—your message, the MVPs, and the market—should align at a critical point of resonance. This is your intended target, where your communication hits the mark and effectively resonates with your audience (see **Figure 61**). When the message's frequency, MVP's natural frequency, and the market's frequency are reaching, your audience can hear you, and you can hear them. You have dialed into understanding their needs, voice, and expectations.

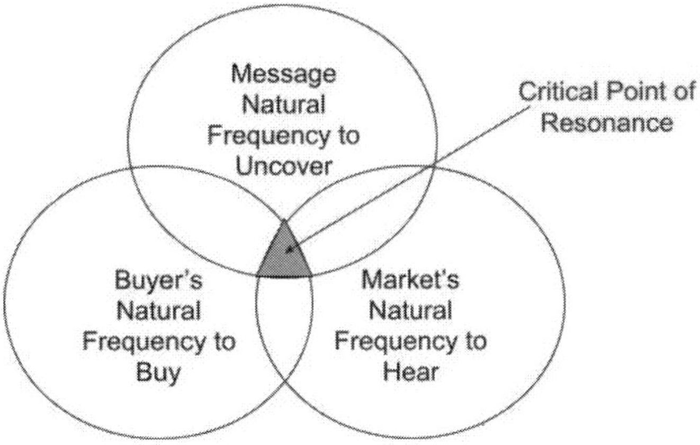

Figure 61. Frequency Resonance

Message-to-Market Match

Your outreach message has a natural frequency that uncovers value for your customer and your product. When you effectively align your message with the market, a vibrational frequency resonates with your target MVPs, allowing them to understand and appreciate your offerings. Conversely, the message may repel those who do not align with your vision. When communication resonates, your target MVP can comprehend it clearly, free from confusion, interference, or distractions. However, if the market message is misaligned, you may find it challenging to connect with your target MVPs, leading to wasted effort in nurturing potential customers who may require convincing that your product is the right fit. This misalignment can decrease filtration on both sides, adding stress and strain to your team and unnecessary noise or friction between your message and your target audience.

It's essential to recognize that buyers tend to purchase based on their desires, not necessarily what they believe they want. Therefore, aligning your philosophy with the buyers' needs wants, and purchasing styles is crucial. Understanding natural frequency is the key to effectively matching the market with your message.

Your organization, clients, and the market all have natural

tendencies that influence how they perceive your solution. Specific word choices and messaging can trigger their attention and facilitate their understanding of your product or service. Your role is to ensure that all three frequencies resonate harmoniously, allowing your products and services to attract the right audience.

When the natural frequency of your MVPs, your outreach message, and the market align, they create a phenomenon known as frequency resonance. Attempting to drive action without ensuring that these three elements are synchronized will inevitably dampen your results. Get this alignment right, and your desired outcomes will resonate with the force needed to excel in your organization.

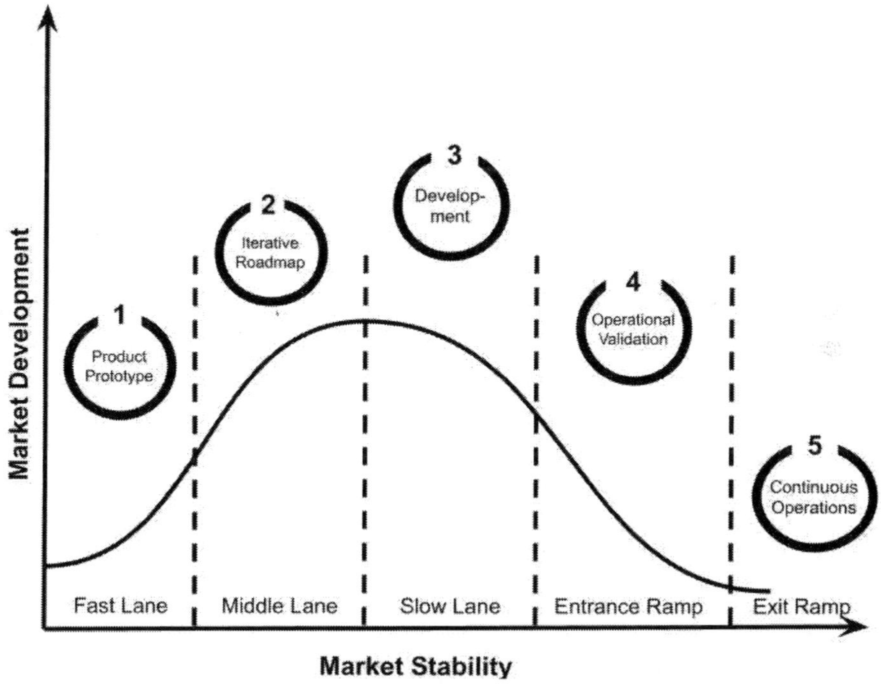

Figure 62. Market Lifecycle Development: MVP-Message Match

Message Development

To effectively engage with your target audience, you must develop your message for each phase of development (see **Figure 61**). Pay attention and recognize that each buyer possesses a natural frequency. This frequency influences how he/she responds to your outreach efforts. You want to mirror this frequency in your message. Understanding this concept is crucial because frequency determines everything. You don't have to recreate the frequency; your MVP will teach you how to communicate with them if you listen and match their frequency. But it's not enough to just mirror their frequency; in your message development, you need to speak to the nuances of a buyer's reaction and engagement within each lifecycle stage. Get to understand and know what your MVP might be experiencing, what memories they have, and what they have done to alleviate their problems. Put yourself in their shoes and speak to them. Let them know you're listening and you get it. This may take some research, but if you listen well, you'll quickly see your MVP has done all the hard work for you; your tasks are to listen, resonate, and give feedback to what you hear so you get it right. Listening and Resonation allow you to build stability and development in your marketing process. Feedback will enable you to fine-tune the details.

To connect meaningfully with your MVPs, you must be aware of the underlying forces at play and the personality types within their organization. Each individual's natural frequency can affect their receptiveness to your message, shaping their decision-making processes and responses. By aligning your communication with these frequencies, you enhance your ability to resonate with your audience and foster stronger connections that lead to successful outcomes.

Market Development

Markets Natural Frequency: Each market possesses a natural frequency that corresponds to the growth stage of your clients' organizations. When the natural frequency of your Most Valuable Prospects (MVPs), your outreach message, and the market's natural frequency align, they create a phenomenon known as frequency resonance. This resonance is crucial; attempting to drive action on any other frequency without ensuring that all three elements are harmonized will inevitably dampen your results. When you achieve this alignment, your desired outcomes will resonate with the force necessary to excel within your organization.

Market-Message Life Cycle: The Market-Message Life Cycle encompasses your company's ability to implement a compelling message that matches your MVPs' journey. The position of your MVPs within their lifecycle will dictate the marketing message, strategy, and approach needed to effectively reach and convert them from potential customers to loyal clients. Segmenting your potential customers and MVPs is essential to tailor your efforts accurately.

In this phase, you will gain insights into the specific needs of the market and its lifecycle, allowing you to craft strategies that resonate deeply and foster strong connections with your audience. Now, let's combine this and explore the bigger picture when connecting your products and message with markets and MVPs.

A. The Roadmap to Success: Fast Lane Innovators in Premium Markets

Fast Lane customers are your ideal collaborators during the prototyping phase. Fast Lane MVPs are the innovators—the forward-thinkers on a mission to stay ahead and set trends. Positioned in the Iterative Prototype stage of the Product Lifecycle, these customers are actively seeking a unique style of communication and premium solutions that can accelerate their goals and vision. They know exactly where they're headed and the tools they need to get there. Pricing seeking value on quality and efficiency. They are prepared to invest at a premium because they recognize the return on saving time and maximizing impact.

These customers are drivers of change, leveraging the Gravitational Force and Creating Innovation personality style. They are the catalysts in their industries, eager to explore new technologies and innovative approaches that disrupt the status quo. They thrive in luxury and specialty markets, where thought leadership, ideas, and unique transformations stand out. Fast Lane Creating Innovators want honesty about your product's capabilities and value straightforwardness in communication. Their endorsement and commitment often pave the way for adoption among the next wave of customers, lending credibility and influence to new solutions. These visionary, creative, and disruptive thinkers view the world as you do. They're eager to embrace the latest and greatest trends, often before others even recognize the opportunity. Those in the fast lane don't just get it—they want it and are ready to buy it with money in hand.

Selling to Fast Lane MVPs is relatively straightforward—they know what they want and commit when they find it. They value providers who genuinely understand their needs, aspirations, and values. By aligning with their vision and addressing their unique requirements, you tap into the premium market, where buyers aren't just looking for solutions—they're looking for community and partnerships that empower their innovation journey. By working with Fast Lane customers, you'll establish thought leadership and position yourself as a market pioneer. Remember: the sprout stage is where leaders emerge.

What to Expect When Prototyping for Fast Laners

1. **Low Market Growth Rate** - True innovation means the market hasn't caught up to your vision yet. Most people won't see the problem—or the solution—the way you and your Fast Lane customers do.

2. **Low Competition** - Competition, defined as "perceived practical alternatives," should be minimal. If your idea feels bold, crazy, or even a little "out there," you're on the right track.

3. **High Costs, Low Pricing Pressure** - Prototyping for Fast Laners is expensive due to the market's infancy. However, these customers are willing to pay a premium for innovation because they understand its long-term value. For them, your premium price is a small investment for a significant return.

Why Fast Laners Matter

Fast Lane innovators are pioneers—they **sense the future before it happens**. Think back to the early days of mobile technology. Ten years ago, when the first mobile companies launched, many people thought they were chasing a pipe dream. However, those early creative innovators pushed the envelope well before smartphones became mainstream and shaped today's mobile boom environment.

Thought Leadership for the Fast Lane

Collaborating with Fast Lane customers establishes your brand as a trailblazer. By solving their core problems and delivering unique value, you're building trust and positioning yourself as the authority in your market. Fast Lane innovators expect to pay a premium, but they view it as an investment, not a cost. They value spending money on the front end to save money, time, and resources on the back end.

The Bottom Line

Prototyping for Fast Laners is about aligning with customers who share your vision and are ready to invest in the future. By focusing on thought leadership, solving core problems, and understanding their willingness to pay for innovation, you set the stage for sustainable growth and market leadership.

B. Middle Lane MVPs - Middle Mass Market Lifecycle

Selling to Middle Lane MVPs is relatively straightforward, but maintaining satisfaction can be challenging, as they are still refining their business acumen and may need time to see the larger picture. Your communication should seek to be more engaging with this group to get their attention, which is focused elsewhere. Working with the middle lane can be rewarding if you provide the thought leadership they need and teach them so they can understand the full potential of your solutions. They are valuable contributors to the product iteration roadmap, as their feedback helps enhance your offerings. When you effectively address their needs and support their journey, they'll continue to buy, seeking solutions that align with their roadmap and progress toward their goals.

Middle Lane: Turning Pain into Progress

Once you've gained traction with fast-lane innovators, it's time to engage with a different type of customer: Middle Lane customers. Unlike Fast Laners, Middle Lane customers are focused on immediate pain relief—they want solutions now, often without concern for the long-term or whether you're a Fortune 500 company. If you can solve their pain effectively, they'll work with you. Middle Lane MVPs are early-stage adopters and the driving force behind the middle market. These clients are navigating the Product Roadmap phase of the product lifecycle, corresponding with their seedling stage of development. They often come to you with a specific business problem or a strategic opportunity they need support to solve, yet they are still exploring what methods and principles will guide their future success. For them, size and scale are less important than your ability to deliver actionable results their team can implement effectively.

Educate and Iterate for Middle Mass Market Customers

Middle Lane customers require a different approach. While Fast Laners see the big picture, Middle Laners are often unaware of the broader issues they face. They only recognize their immediate pain, making it necessary to address their challenges one problem at a time. Middle Lane MVPs require close collaboration and guidance to identify and address their core challenges. They lean on the steady, results-oriented Strong Nuclear Forces within your organization to help produce outcomes in the early stages, and they need Electromagnetic forces to document their growth and capture results along the way. Though not premium market buyers, they represent a crucial segment that often becomes loyal customers if you help them achieve consistent progress. For those who may not advance to the fast lane, these clients still offer substantial value, as they are poised to become your next-highest value buyers. Here's how you engage them:

1. Build on Your Thought Leadership - Your work in the fast lane establishes your credibility. Now, Middle Laners need education and proof that you understand their pain and can solve it.
2. Iterative Solutions - Break down their core problem into smaller, manageable micro-problems. Offer solutions step by step, creating

a roadmap that gradually introduces your full product or service suite.

3. Document Results - Provide testimonials, case studies, and measurable evidence of your effectiveness. This builds trust and attracts more customers with similar pain points.

What Happens Next?

Engaging Middle Lane customers has measurable effects:

- **Market Growth Increases** - Your proven solutions and established thought leadership start to drive awareness, drawing attention from other customers who share the same pain.
- **Pricing Pressure Decreases** - As your credibility grows, customers are willing to pay more for your solutions. Though initially solving smaller problems, the cumulative value justifies premium pricing.
- **Few Competitors** - You still face limited competition because your solutions are unique and tied to your thought leadership.

Monetizing the Middle Lane

Selling to Middle Lane MVPs is relatively straightforward, but maintaining satisfaction can be challenging, as they are still refining their business acumen and may need time to see the larger picture. Selectively engaging with this group can be rewarding if you provide the thought leadership they need to understand the full potential of your solutions. They are also valuable contributors to product iteration, as their feedback helps enhance your offerings. When you effectively address their needs and support their journey, they'll continue to buy, seeking solutions that align with their roadmap and progress toward their goals. In this stage, you're generating revenue in two ways:

1. **Serving Fast Laners** - Continue providing advanced solutions to Fast Laners who already see the value in your full product or service.
2. **Iterating for Middle Laners** - Sell components of your core

product to Middle Laners, solving their immediate problems while gradually introducing them to the broader value of your offerings.

While Middle Laners may not yet purchase your full product, the iterative model allows you to sustain growth. Charging a premium for smaller components positions your brand as both accessible and valuable.

The Seedling Stage of Business Development

By engaging with Middle Lane customers, your business transitions from the sprout stage to the seedling stage. At this point, you are communicating topics that align with:

- Establishing your business structure.
- Building your team and operational capacity.
- Documenting proof of success and solving customer pains.
- Laying the groundwork for a productized suite of services.

Recap

Here's how the roadmap evolves in the middle lane:

1. **Thought Leadership:** Use your established authority to educate Middle Laners and address their immediate needs.
2. **Iterative Roadmap:** Solve micro-problems step by step while building toward a comprehensive product suite.
3. **Proven Results:** Provide tangible evidence of your solutions to build trust and attract more customers.
4. **Sustained Growth:** Serve Fast Laners with advanced solutions while monetizing Middle Laners with iterative offerings.

The middle lane is where you bridge innovation and scalability, setting the stage for long-term success.

C. Slow Lane: Scaling for Sustainable Mass Market Growth

When your business reaches the slow lane, it's time to communicate on how to scale and grow into the maturing stage of the business market lifecycle. The Slow Lane consists of early majority customers, often referred to as the mass market, who prefer stability over risk. In this developmental lifecycle phase, akin to the Sapling stage, these practical buyers make up the largest segment of potential customers. They view risk as a waste of both time and money, which means they require the steady influence of Strong Nuclear Forces to help produce reliable results, Electromagnetic Forces to document results, and Gravitational Forces to innovate and create new products. Their interactions are largely shaped by Connecting Communicator personality types, who facilitate effective communication and help maintain consistent outcomes.

Interestingly, Slow Lane MVPs appreciate market competition, as it enables them to compare and verify their purchases against established, proven options. Without clear differentiation, however, they tend to prioritize price and discounts, making them sensitive to cost factors. Their purchasing decisions are heavily influenced by the overall stability of the product, alongside the cost of goods. By understanding the motivations and preferences of Slow Lane MVPs, you can tailor your approach to effectively meet their needs and secure their loyalty.

Systematize and Scale

By now, you've established a profitable system. You've proven that your solution addresses a core customer pain, and your ability to **sell, service, and deliver** is systematized and efficient. Once Slow Lane MVPs identify a viable solution, they exhibit remarkable loyalty and seek to ensure that the product aligns with the needs of their entire organization or industry. They invest considerable time in researching and validating potential solutions, prioritizing the product's quality, the supporting infrastructure, and the reliability of the service. This stage of product-solution validation is critical, as it safeguards their interests and ensures that everything is executed correctly without exposing them to unnecessary risk. Customers are willing to pay a premium for a **comprehensive solution**, and your business is reaping the rewards:

- **Profit margins increase** as your processes mature.
- Your internal needs are met, ensuring sustained operations.
- Your **thought leadership** amplifies your credibility, making customers confident in paying for your expertise.

Prepare for Competitors

As your success grows, so does your visibility. Expect **competitors** to emerge, trying to claim a piece of your market share. However:

- You're **unshaken** because your foundation is strong.
- Years of preparation and strategic growth have given you a deep-rooted advantage that competitors can't replicate.
- To them, you may seem like an "overnight success," but your established systems, expertise, and customer loyalty prove otherwise.

Leverage Your Leadership

Competitors may seek partnerships or guidance from you. This is an opportunity to:

- Qualify potential collaborators based on **aligned vision and values**.
- Offer **affiliate opportunities** where appropriate, sharing profits while maintaining control.
- Keep your **standards high** to ensure only the right partnerships enhance your brand.

Benefits of Scale

In the slow lane, you're enjoying the benefits of scale:

- **Pricing pressure decreases** as customers see the value of your offerings.

- Your established reputation lets you charge premium rates while facing less resistance.

Revenue streams diversify as you introduce multiple ways to deliver your products or services, such as: tiered pricing models, specialized offerings for different customer segments, and new delivery methods or formats. With the market booming, competition won't undercut your success because your processes, reputation, and customer trust are unshakable.

Positive Returns and New Ventures

This is the stage where your business unit generates consistent **positive returns**. With a steady foundation, it's time to:

- Launch **new business units** that will go through their own lifecycle stages.
- Use your established systems and insights to replicate success across new ventures.

Recap: The Slow Lane

The slow lane is about scaling, sustaining, and thriving:

1. **Profitability**: Streamlined systems and proven solutions drive profits.
2. **Competitor Advantage**: The deep reservoir of thought leadership built ensures you remain unthreatened by new players.
3. **Diversified Income**: Multiple offerings and delivery methods maximize returns.
4. **Expansion**: Launch new ventures and business units to build on your success.

By focusing on strategic scaling and leveraging your position, the slow lane becomes your runway to long-term growth and impact.

D. Avoiding the Commodities Market: Navigating the Entrance Ramp

The final stage in a business lifecycle is the **entrance ramp**—a stage where businesses risk stagnation or decline. The Entrance Ramp MVPs are situated in the Aging Tree lifecycle developmental phase, characterized by their presence in the Operational Validation Product Lifecycle stage. These customers primarily maintain the status quo, with their business practices becoming the industry norm. They often adopt a race-to-the-bottom strategy, treating products as commodities rather than unique offerings. Clients in this category tend to be large companies focused on sustaining their operations, where effective communication, timely product completion, and reliability are crucial. Given their risk-averse nature, In the early stages, they rely on Strong Nuclear Forces, Weak Nuclear Fores, and Electromagnetic Forces to ensure consistent results. As it ages into dying, Electromagnetic Forces will continue to document and process the sale of community-based content in the later part of this phase. The reality is that if you can recognize the signs, you can avoid their early onset. As you can see, innovation is not needed at this phase of market development.

The Reality of the Entrance Ramp

1. **Customer Profile**: Entrance ramp clients are bargain hunters. They prioritize low cost over innovation and prefer commodities. They're not looking for premium solutions or cutting-edge innovations—they just want "good enough."

In terms of pricing, vendors and service providers are expected to provide quotes or bid on contracts, leading to a competitive environment where prices are driven down. This market segment typically makes purchases only when products are fully mature and well-established. Entrance Ramp MVPs favor highly developed offerings and prefer to buy from reputable companies with a strong track record. They often seek all-inclusive package deals, where every aspect is bundled and accounted for in the price.

2. **Crowded Market**: This space is saturated with competitors. It's

crowded at the bottom, with "pick-me" businesses all vying for attention by lowering prices.

To attract this market, building a team of strong connecting communicator personality styles who can connect the dots and communicate value is your strategy to compete against a market that tends to resist new products and innovations and prefers to engage with businesses and brands that boast a documented history of success. Introducing new products to this market requires not only a change in mindset but also a shift in behaviors. Entrance Ramp MVPs focus on building extensive databases of vendors, contractors, subcontractors, and suppliers, which they can qualify, validate, and leverage as their needs evolve.

3. **Stalled Growth**: Market growth slows significantly as problems have been solved and innovation stagnates. Pricing pressure mounts, forcing businesses to lower prices and reduce profit margins.

To succeed in this market, it's essential to establish strong leadership, achieve economies of scale, and dominate key distribution channels. By understanding the preferences and behaviors of Exit/Entrance Ramp MVPs, you can tailor your approach to meet their expectations and secure a foothold in this competitive landscape.

Strategies to Sustain Profitability

Even in this stage, with the right approach, a business can remain profitable:

1. **Repackage Offerings**: Adapt your product or service to fit this market segment. Think simplified or bundled options that match commodity-level expectations.
2. **Introduce Partnerships**: Explore sponsorships, collaborations, or affiliate models to create additional revenue streams and reduce costs.
3. **Segment Appropriately**: Offer products specifically tailored to the entrance ramp clients without diluting your core offerings for

premium customers.

The Value of Doing It Right

If you've followed the proper sequence:

1. **Sprout It**: Prototyped for innovators in the fast lane.
2. **Seedling It**: Iterated for Middle Lane customers.
3. **Sapling It**: Systematized and scaled for slow-lane clients.
4. **Grown It:** Smartly adapted for the exit ramp buyers.

Even in the dying phase, your business can remain **lucrative** by strategically serving this segment while maintaining your broader market presence.

Key Indicators of Success

Use these metrics to assess your strategy at every stage:

- **Market Growth Rate**: Stay ahead of stagnation by continuously innovating and adapting.
- **Competition**: Monitor how crowded the space is and differentiate your offerings.
- **Pricing Pressure**: Resist unnecessary price cuts by delivering clear value.
- **Net Cash Flow**: Ensure your cash flow supports future investments and mitigates risks.

CHAPTER 23: Business Execution

Now, let's delve into the critical phase of business execution. This stage emphasizes the importance of having the right combination of messaging, organizational development, and stability to support the various stages of your product and market effectively. It's about balancing (1) development and (2) stability represented by the fuel injector pump leading to the engine.

The Fuel Injector Pump Represents Business Execution

Central to this process is the Fuel Injector Pump (see **Figure 63**), which actively removes micro-debris from your fuel. The fuel injector infuses the engine with new energy as needed.

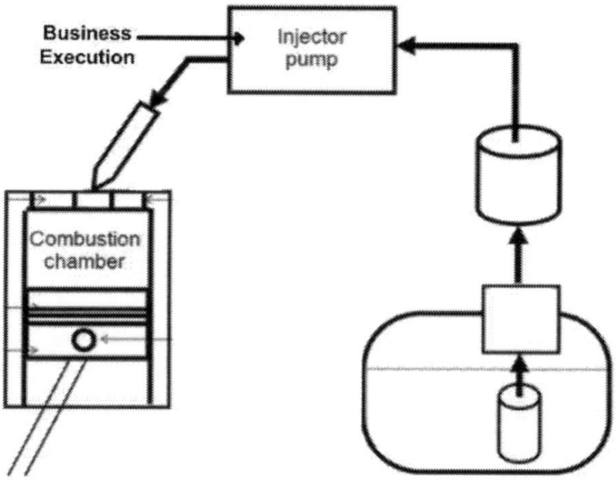

Figure 63. Mechanical Fuel System: Fuel Injector Represents Business Execution

The synergy of all four components of the Mechanical Fuel System is essential, as they work together to provide the energy that powers your organization's engine. By aligning your messaging and maintaining organizational stability, you will enhance your execution capabilities, paving the way for sustained success and growth.

Business Execution Lifecycle

The goal of the Business Execution Lifecycle is diversity. You want to create a healthy mix of the right amount of support and development for the business model, products, and markets at this stage of the lifecycle for the entire organization. If the organization is in startup mode, the entire organization should focus on supporting and developing the organization. If it's in the decline stage, a different type of support is needed to nurture new business units and maintain maturing business.

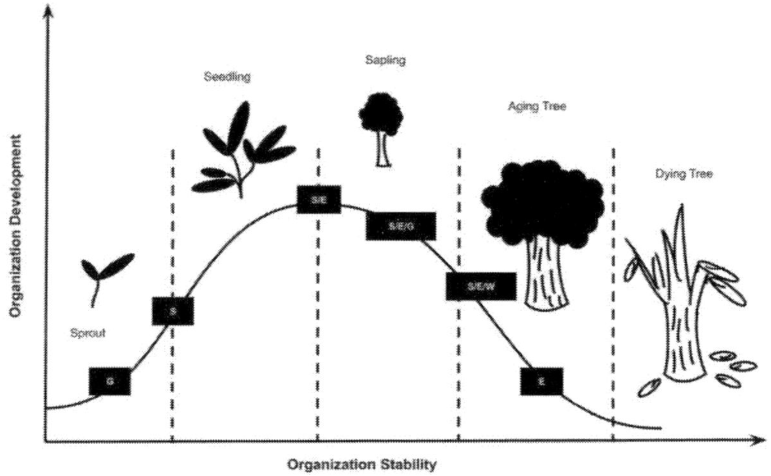

Figure 64. Business Execution Lifecycle

Stability-Development Balance

When selling to people in the fast lane, it's crucial for your company to be in the sprout state of growth and focused on innovation and future advancements. Gravitational Force is a creative force, so it's important to have an innovative company culture to bring new ideas to life. If your product or service is groundbreaking and the leader is passionate about it, then your company is naturally innovative. This sprout stage must take as much time as needed to discover important new ideas, develop the foundation for leading thinking, and truly understand the product-market fit.

It is not necessary to create strict structures, processes, or regulations that hinder creativity. At this stage, the organization requires a creative leader who is extremely innovative and has a defined vision for the product-market fit and the direction to steer the organization.

The organization needs to run cost-effectively, cutting down on unnecessary expenses. Any investments should be directed towards creating prototypes. The team must closely control any unwanted results and chaos while focusing on developing innovative ideas and strategies.

The goal of the Business Execution Lifecycle is to operate with stability and development by putting everything from previous chapters together so the business can regenerate itself and continue to mature, duplicate itself, create new products and services, reach new markets, and eventually add new business models that support themselves while shedding any old or excess weight pulling the system down or draining resources, as seen in F**igure 65**. Let's dive deeper to learn how the stability-developmental should executed.

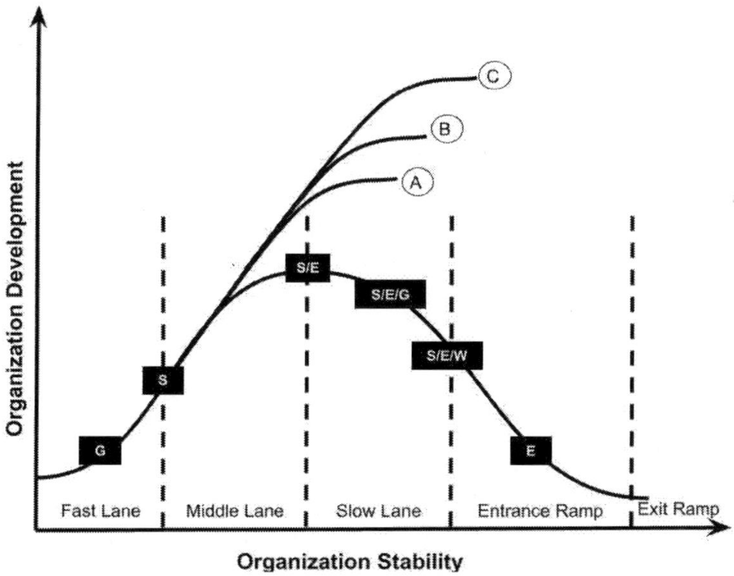

Figure 65. Stability-Development Lifecycle Renewal

Sprout Development

In this sprout stage of growth, your customers and clients are primed and ready to buy. This phase is characterized by Creative Innovators who are eager to engage with you. Think of the fast lane as having the green light and money in hand; they recognize you as the expert who can provide the solution they seek. These customers are curators of premium and cutting-edge products and services. Leverage your Gravitational Forces to attract these Creative Innovators effectively, positioning your offerings as the ideal fit for their needs.

When selling to clients in the fast lane, it's essential for your company to be in the Sprout phase of growth, focusing on innovation and future advancements. This developmental stage is about gathering the necessary resources and materials to set the foundation for future success. Your organization should explore its surroundings and strategically position itself for what's to come. Embracing Gravitational Force as a creative driver is vital; fostering an innovative company culture is key to bringing new ideas to life. If your product or service is groundbreaking and the organizational leader is passionate about it, your company is on the path to becoming naturally innovative.

In this phase, your client's organization will benefit from a Creating Innovator personality type that helps develop the solutions they need for growth. Your organization must leverage Strong Nuclear Forces to effectively "guard" and "implement" your ideas. It's crucial to allow ample time for discovering important new concepts, developing a foundation for leading thought, and truly understanding the product-market fit.

To ensure efficiency, your organization should operate cost-effectively, minimizing unnecessary expenses. Any investments made should prioritize the creation of product prototypes. The team must closely monitor any unwanted results or chaos while remaining focused on developing innovative ideas and strategies.

Fast Lane clients are typically filled with Creating Innovators during this sprout phase. They are eager to hear ideas, strategies, and solutions that can benefit the entire organization.

Seedling Development

This seedling stage represents early growth. Here, you'll need your Strong Nuclear Forces to serve as Implementing Guardians, producing the necessary results as the organization gears up for expansion. Customers in this group recognize their need for assistance but may not see it as urgent or require only a portion of the solution you offer. They often need help implementing, hiring teams, and utilizing tools and systems to carry out their tasks effectively.

When executing the Seedling Phase, it's essential that your MVP customers are also in the seedling development stage. During this phase, organizations focus on protecting their gains while nurturing the client, team, and product. The seedling phase emphasizes iterative development, where the primary goal is to deepen understanding and achieve value-added results that propel progress. If your company is in this stage, you should similarly prioritize the iterative development of your products.

These MVPs fall into the Middle Lane, and they require Weak Nuclear Forces to help "nurture" and "unify" key processes and produce meaningful work. A results-driven company culture is crucial at this stage. Your client's organization needs you to provide nurturing and unifying support. To facilitate this, ensure you have client-facing and team-facing supervisors who can focus the teams effectively. Your product should assist them in their business journey, helping to educate and bridge their understanding of the challenges they face versus the solutions you provide.

By concentrating solely on the MVP's immediate needs, avoid the temptation to create overly complex products. Collaborate closely with your clients, as they will communicate their requirements. Listening and providing adequate education are vital during this development stage, as they will need guidance, direction, and confirmation.

Take the necessary time in this seedling phase to allow your clients to discover their values, quality assumptions, and ideas. This is the time to develop a strong foundation for innovative thinking and to understand the product-market fit thoroughly. Avoid imposing rigid structures or regulations that may stifle creativity. Instead, seek a creative leader who is innovative and possesses a clear vision for aligning the product-market fit with the organization's direction.

Your organization should operate cost-effectively, minimizing unnecessary expenses. Any investments should focus on creating product prototypes. Your team must manage any unwanted results and chaos while remaining dedicated to developing innovative ideas and strategies.

Middle Lane clients are often represented by Implementing Guardians in the seedling phase of their organization. They are ready to get to work. They need to hear about implementation assistance, team hiring, and the best tools and systems to help them execute tasks and establish their subsystems effectively.

Sapling Development

In the execution phase, organizations must focus their communication on providing valuable advice, managing systems effectively, and preparing for scalability to achieve mass exposure. The Sapling phase of developmental growth centers on growth and scaling. During this phase, your marketing efforts should aim at increasing profit margins and revenue, capturing market share, documenting processes, and managing client relationships and data. Your actions need to emphasize visibility and authority, leveraging Strong Forces to "guard" and "implement" results and Electromagnetic Forces to drive growth while ensuring the team can effectively "connect" and "communicate" value through data analysis. In the middle of the phase development, this shifts to include Gravitational Forces to bring in new ideas, "create" new products, and "innovate" as it scales. As the sapling stage develops into the end of the phase, it's time to stabilize. Gravitational Forces shift into Weak Nuclear Forces as "nurturing" and "unifying" the teams becomes the focus.

Your MVPs are in the slow lane, concentrating on growth. They will find the most value in identifying opportunities for expansion and forging strategic partnerships. If your company is in the seedling phase, your focus should also be on standardizing, simplifying, and managing resources. By this stage, you should be at your peak in terms of lifecycle development, competing effectively in your market space with a differentiated presence in your field. In this phase, you're providing shade and nutrients for other businesses and developing additional products and services to support your ecosystem.

Slow Lane clients need a combination of Connecting Communicators and Guarding Implementors in the sapling phase of their organizations. They require insights into industry best practices, competitive guidance, and support for managers and supervisors. Additionally, they seek to expand benefits that will help retain talent across the entire organization.

In the sapling phase of growth, your messaging matures. This is where Connecting Communicators become essential to provide more information. Customers in the slow lane are at a red light; they need you to connect the dots for them. They may not be aware of the challenges they face and require education to illuminate the voids and issues that need addressing. Your focus should be on nurturing relationships, establishing systems, and providing management solutions that help retain talent within their organizations.

Aging Tree Development

In this execution phase, organizations must prioritize communication centered around the maintenance and stabilization of systems while preparing to ride the wave of success. If you've executed everything correctly, this wave could sustain itself for years. Your focus should include continual monitoring and upkeep. The Aging phase of developmental growth emphasizes sustainability.

During this phase, your marketing strategy should revolve around consistency, corrections, and dependability. This involves ongoing monitoring, reputation management, maintaining quality standards, and actively engaging with customer feedback loops.

To navigate this phase effectively, you need a trio of forces: Strong Nuclear, Weak Nuclear, and Electromagnetic Forces. Each of these forces plays a crucial role in implementing work, nurturing the product, and communicating needs, all aimed at maintaining the status quo of what has been successful. Your MVPs are in the entrance lanes, navigating the complexities of thriving and declining as they approach the end of their lifecycle. Innovation may not be a priority at this stage. Instead, data will dictate when it's time to retire certain products or services with Electromagnetic Force in control. The information gathered can become valuable resources, disseminated through websites, webinars, and conferences to fuel future innovation and rebirth.

Dying Development

In this dying developmental phase, your messaging resonates best with communicators. Buyers in this segment are browsing the market and staying updated with the latest offerings. They tend to build extensive vendor databases for future reference and exhibit a risk-averse mindset. While they may seek innovative ideas to upgrade or enhance existing systems, products, and services, they require clear communication, organization, and legacy systems to support their needs effectively.

Exit ramp clients are represented by Connecting Communicators. As your organization enters the Aging Tree phase, these clients will require communication focused on maintaining their systems and updating documentation to support legacy systems and subsystems.

SECTION IV: The Right Execution for Speed

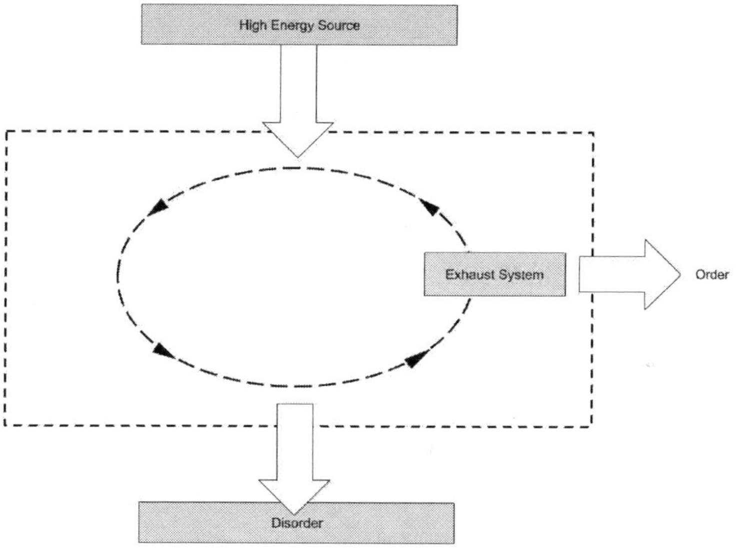

Figure 66. Mindset Mechanics Interface: Exhaust System

CHAPTER 24: Align for Speed of Implementation

The *Exhaust System* is a crucial component that measures the overall condition of an organization's output as energy flows through its various processes (see **Figure 67**). Just as a vehicle's exhaust system has evolved over the years to enhance performance, an organization's exhaust system can provide insights into its operational efficiency. The exhaust system guides depleted gases away from the engine, playing a vital role in determining performance, fuel economy, power, material quality, and entropy (or disorder). By monitoring this system, you gain a clearer understanding of the health of your organization and can identify areas of disorder and chaos.

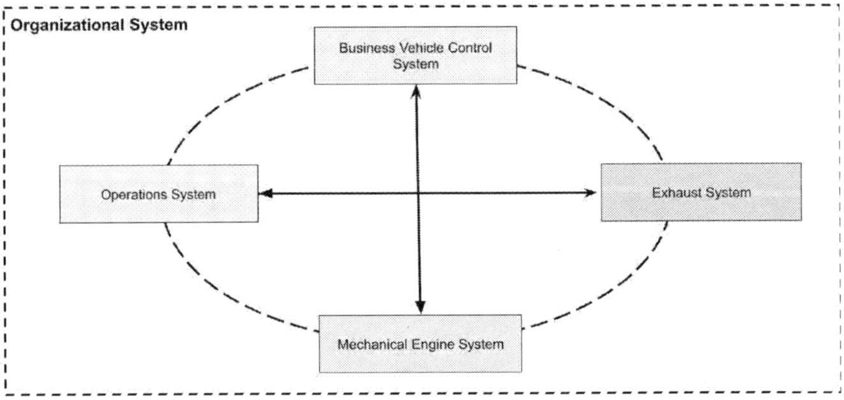

Figure 67. The Exhaust System

Many functional mechanical operators may only focus on sounds or speed when evaluating exhaust systems, overlooking other critical factors such as weight reduction, torque, horsepower, and the overall lifespan of vehicle performance. The exhaust system is pivotal in assessing the health of your business vehicle, as it reveals whether critical subsystems are aligned or misaligned. If misalignment occurs, your business vehicle requires mechanical servicing to identify the root causes of undesired output.

The exhaust system serves several important functions to help determine the overall health of the organization:

1. It helps determine if critical business vehicle subsystems are aligned or misaligned. If the business vehicle is misaligned, mechanical servicing is needed to check the cause of the undesired output.
    a. It aligns organizational mass—focusing on internal culture and external structure.
    b. It aligns business velocity—streamlining the decision-making and implementation process.
    c. It aligns business acceleration—bringing together various organizational forces.

2. To maintain optimal performance, the exhaust system also aids mechanical operators in troubleshooting issues.
    a. The first step is a *Mindset Check*, involving a visual, eight-point preventative maintenance inspection to identify potential leaks within the system.
    b. Following this, a *Mechanics Check* connects the entire business vehicle to a diagnostic machine that collects data, monitors performance, and assesses whether any subsystems are at risk of failure.

Go Slow to Go Fast

You've done all the groundwork and now grasp the principle of "go slow to go fast." However, if you find yourself or your organization stagnating or moving at a sluggish pace, no amount of desire will help you speed up in the way you might expect. This stagnation often results from not overcoming the forces of inertia that keep you grounded. If you or your organization are heading in the wrong direction, you need to slow down, pivot, and redirect your efforts. In this case, you're experiencing momentum, and it's essential to apply inertial forces to halt your current trajectory, reassess your goals, and realign your direction.

To foster momentum and acceleration, we must first ensure that all elements are properly aligned. Let's integrate everything you've learned in Mindset Mechanics and prepare to create harmony among the components that drive organizational momentum and acceleration. Understanding how forces like inertia affect motion is crucial for this process.

Organizational Momentum

Momentum is exactly what you want to work for you in your organization. The equation for momentum is given as *mass* times *velocity*.

$$p \text{ (momentum)} = \text{mass} \left(\frac{\Delta x \text{ (displacement)}}{\Delta t \text{ (time)}} \right)$$

Figure 68. Business Momentum is mass times displacement over time.

To build organizational momentum, the first step is to accumulate your total mass, which we refer to as organizational mass multiplied by business velocity.

As discussed in chapters 8 and 9, organizational mass consists of two components: internal mass and external mass. Internal mass encompasses your organizational culture, including the alignment of your personal and organizational vision, values, goals, and mission. On the other hand, external mass refers to your organization's structural design and function.

In addition to mass, calculating organizational momentum requires understanding velocity, which we define as business velocity. Business velocity is the ratio of displacement over time. Displacement represents the decisions and actions you take that differentiate your organization's current state—often marked by pain, frustration, or low results—from the future state you aspire to achieve. We will explore business velocity in greater detail in the next chapter.

Organizational Acceleration

Momentum is great, but acceleration takes things to the next level. Acceleration is where the real excitement happens within your organization. True organization is achieved when you have talented individuals collaborating with you to amplify your results.

Organizational acceleration is defined as the sum of organizational forces divided by the sum of organizational mass. The formula for acceleration (see **Figure 61**) highlights the importance of not only having the right resources (Forces) and culture (mass) but also harnessing the collective strength of your team to propel your organization forward at an impressive pace.

$$\text{a (acceleration)} = \left(\frac{\sum \text{F (Force)}}{\sum \text{m (mass)}} \right)$$

Figure 69. Business Acceleration is the sum of the forces over total mass.

Together, high-performance teams, a strong organizational culture, and an effective structure enable business acceleration. Once every component is aligned, you're primed to align (or realign) your subsystems, build (or rebuild) momentum, and accelerate confidently, knowing you're set for a smooth journey ahead.

Building Momentum and Acceleration

The secret to building the momentum and acceleration that moves you toward your desired location with speed is found in Newton's Laws of Motion. In *Principia Mathematica Philosophiae Naturalis*, Sir Isaac Newton developed theories of motion in the 17th century at the age of 23. These three laws revolutionized science and etched Newton's name onto the pages of history forever. Now, let's learn how physics helps move your organization forward.

Newton's Laws of Motion

The key to creating the momentum and acceleration needed to propel you toward your desired destination lies in Newton's Laws of Motion. In *Principia Mathematica Philosophiae Naturalis*, Sir Isaac Newton introduced these groundbreaking theories of motion in the 17th century at just 23 years old. His three laws transformed our understanding of physics and solidified his legacy in history. Now, let's explore how these principles can help propel your organization forward.

Let's explore the concept of motion in periodic systems to better understand how organizations progress. A periodic system operates in cycles, exhibiting repetitive motion along a circular path. This is analogous to how your business functions through various cycles—daily, weekly, monthly, and even annual reports—all of which help you manage operations effectively. Periodic systems serve as an excellent model for comprehending the cycles of business in motion.

Inertia can be illustrated through a simple pendulum, which consists of a mass (the bob) hanging motionless. If left undisturbed, the pendulum will remain still indefinitely. However, once a force is applied under the right conditions, the pendulum can swing or cycle continuously. In this scenario, inertia keeps the pendulum at rest, while momentum and acceleration are what keeps it moving. Similarly, your business experiences inertia, momentum, and acceleration. When operations come to a standstill, they will remain stagnant unless an action is taken to initiate movement. If your business is moving in the wrong direction, it will continue along that path unless a corrective force is applied.

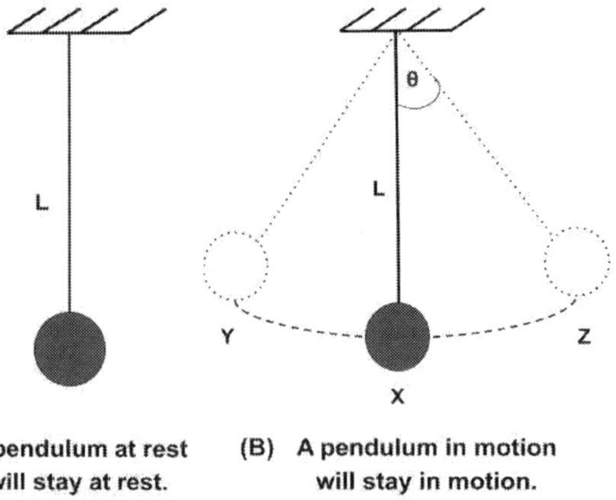

Figure 70. Newton's First Law of Motion

To set the pendulum in motion, you must overcome the "staying force" of inertia. In practice, this means pulling the pendulum back and releasing it, allowing it to swing freely. According to the law of motion, a simple pendulum will move periodically, displaying oscillating momentum. Under ideal conditions, momentum should sustain the pendulum's swing indefinitely. However, what happens if you pull the pendulum further back? It will swing higher, and its motion will last longer. Essentially, the farther you aim to propel your organization, the more torsional force you must apply to overcome inertia.

To get a simple pendulum to move, you need to overcome the "staying force" of inertia. In our case, you give it a pull and let go. Now, the pendulum swings freely. The law shows a simple pendulum moves periodically with oscillating motion or momentum. Momentum should keep the pendulum in swinging motion forever, under ideal conditions. What happens when you pull the pendulum farther back? The pendulum moves higher than before, and its motion lasts even longer. In essence, the farther you need to take your organization, the more torsional force you need to apply to overcome inertia.

Torsional force is the energy required to overcome inertia in organizations that are either stagnant or moving slowly. It also serves to slow down or redirect organizations that are currently gaining momentum in the wrong direction.

Newton's First Law of Motion

According to Newton's First Law of Motion, an object in motion remains in motion, and an object at rest remains at rest unless acted upon by an external force. This principle highlights inertia—a friction force that resists change based on the object's mass. In the case of our pendulum, its inertia dictates its tendency to either maintain its state of rest or its state of motion under certain conditions.

While your organization isn't a simple pendulum, it behaves similarly. Just like the pendulum, your organization will continue its current trajectory unless an external force is applied. If your business is spiraling out of control—losing money, experiencing high turnover, or feeling stuck—it will remain in that state unless you take decisive action to change course.

The greater the mass of your organization, the more significant the forces of inertia that must be overcome. More mass translates to more inertia, necessitating greater effort to initiate change and redirect momentum. Essentially, you will need to apply more torque to get your organization moving in the right direction.

If your organization is experiencing high-speed momentum, it's akin to trying to halt a runaway freight train heading down the wrong track. Although challenging, with sufficient time and space, it is possible to slow it down. The majority of functional mechanical operators—around 98%—struggle with the adverse effects of inertia and momentum, finding themselves in a constant battle for survival against the forces of inertia. In contrast, the remaining 2% of effective operators thrive on the positive side of inertia and momentum, turning business into a game where they leverage Newton's First Law to their advantage.

Newton's Second Law of Motion

Moving on to Newton's Second Law of Motion, we explore the intricate relationship between mass, acceleration, and net forces. This law states that the net force acting on an object is responsible for its acceleration, summarized by the formula $\sum F_{net} = ma$.

In this equation, F_{net} represents the net force applied to the object, m is the mass of the object, and a is its acceleration.

To understand this concept in the context of our pendulum, we can analyze the forces at play. The downward force—due to gravity—acts on the pendulum bob, while the upward force is exerted by the tension in the string, which pulls back on the pendulum rod. By equating these two forces, we can determine the net force acting on the pendulum, which in turn allows us to calculate its acceleration.

This principle is crucial not only for understanding the mechanics of the pendulum but also for applying these concepts to organizational dynamics. Just as the pendulum experiences forces that determine its motion, your organization must contend with various forces that influence its trajectory. Recognizing and managing these forces effectively can lead to better outcomes and greater organizational acceleration.

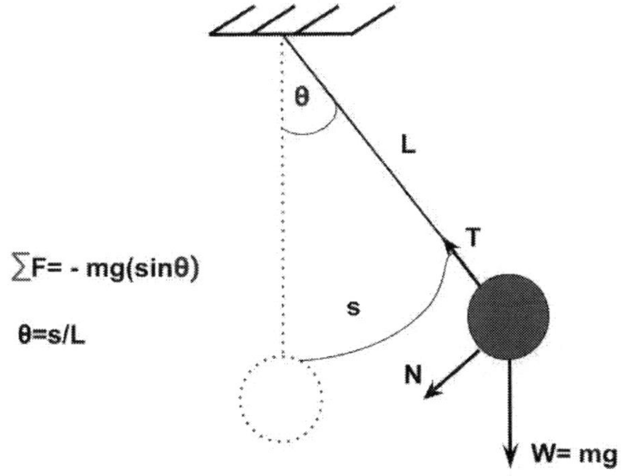

Figure 71. Newton's Second Law of Motion: Net forces acting on a simple pendulum in a cycle.

To increase organizational speed, it's essential to cultivate a high-performance team equipped with the right organizational forces. This high-performance team takes the momentum you've created and amplifies your results. Rather than making one decision and implementing it over time, your team is empowered to make multiple decisions and execute various processes simultaneously. This approach fosters perpetual motion on a larger scale, significantly accelerating your outcomes.

The success of this strategy lies in the shared journey of each team member. Each individual manages a portion of the distance toward the desired destination, contributing to the collective effort. Newton's Second Law of Motion highlights that pendulums with shorter distances oscillate with higher frequencies, a concept that parallels how nimble teams can adapt and respond quickly to challenges. Just as a pendulum swings more rapidly with less distance, a well-aligned, high-performance team can achieve results faster and more efficiently.

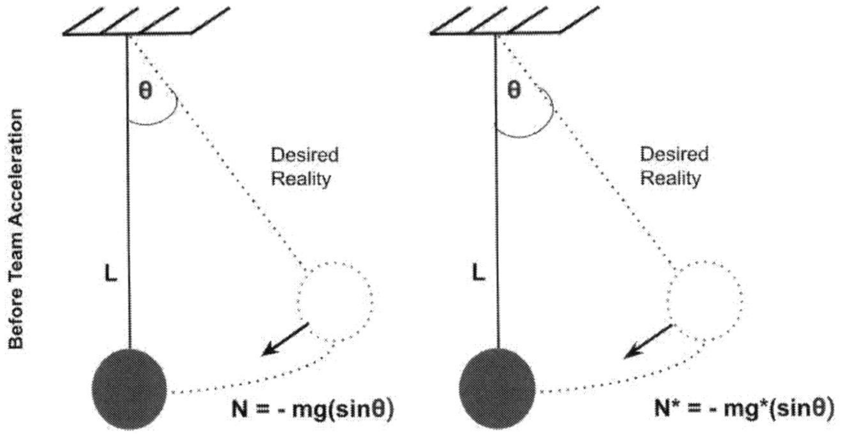

Figure 72. Newton's Second Law of Motion for Simple Pendulum: Increase acceleration due to gravity to oscillate faster.

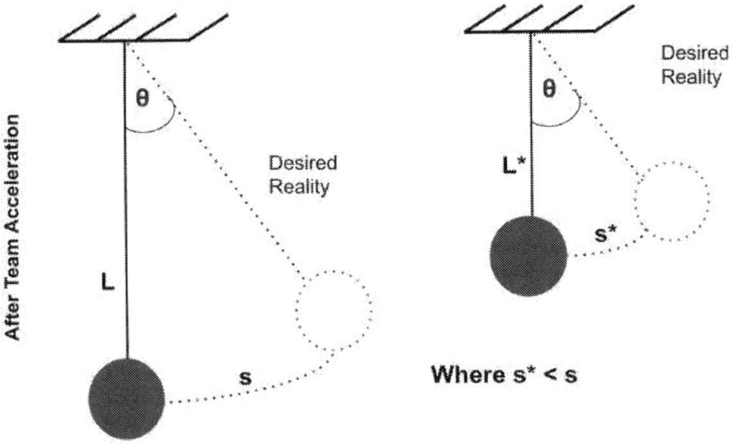

Figure 73. Newton's Second Law of Motion for Simple Pendulum: Shorten the length, L, of the rod to oscillate faster.

Newton's Third Law of Motion

Newton's Third Law states that for every action, there is an equal and opposite reaction. This law highlights the gravitational force that exists between two masses, and in the context of our simple pendulum, it illustrates how vertical tension in the rod holds the bob mass in position while horizontal acceleration from downward forces causes the rod to move side to side.

This principle emphasizes that all organizations experience tension forces. Tension can be beneficial. In our pendulum, tension pulls upward and downward in equal and opposite directions. When you make a change in one area of your organization, there is an equal and opposite reaction, creating a tension that needs to be managed.

In healthy organizations, balanced tension often goes unnoticed because it transforms into mechanical energy at work—specifically, kinetic and potential energy. This mechanical energy represents the actual and potential work your team invests in, driving your organization forward. Essentially, Newton's Third Law illustrates how tension, when effectively harnessed, enables organizational mass and forces to collaborate, allowing the pendulum—and your organization—to accelerate.

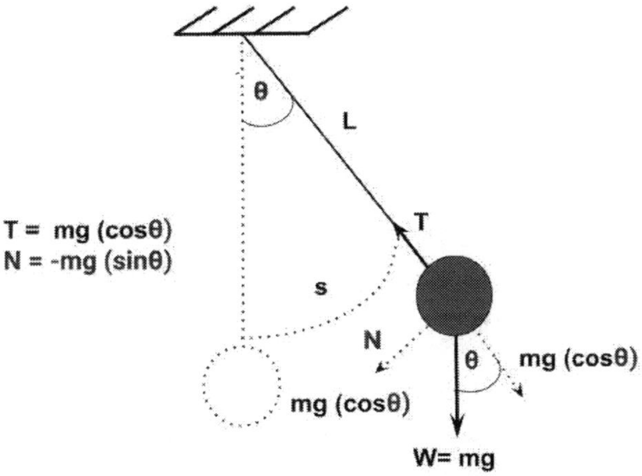

Figure 74. Newton's Third Law of Motion for Simple Pendulum: All forces acting on the bob mass have equal and opposite reacting forces.

Let's examine a practical example of Newton's Laws of Motion in action. Imagine you're eager to increase your revenue quickly—a goal you've long pursued. After exploring various revenue streams, you've decided to change your business model and finally found one that promises to boost momentum. To facilitate this transition, you hire a consultant to implement key components of the new model.

However, as the push for increased revenue takes shape, your business culture begins to experience tension, shifting towards a more traditional organizational structure. Now, there's an expectation for your team to be physically present in the office to achieve financial returns. This represents the equal and opposite reaction at play. What went wrong? A core aspect of the new business model relies on cold calling and television advertising, which contradicts the flexibility you initially envisioned.

Now, you're faced with a dilemma. You've already invested thousands into this consulting firm; momentum has picked up, and the team is engaged. Yet, the culture of your organization is veering away from its original vision and values. You aimed for a flexible work environment, but now some team members are tethered to their phones, pouring their energy into a declining market segment that demands more labor and less creativity.

It's crucial to manage this reaction. The consultant you brought on board may not have fully grasped your organization's structure, design, and function. Increasing sales should not translate into heightened insurance, rent, or labor costs, nor should it diminish team vitality. You've invested significant effort into your new plans, and if you fail to respond to this structural shift, Newton's Third Law of Motion will come into play, accelerating your company's culture, personnel, and processes in a direction you never intended.

Organizational Alignment Increases Momentum

So, how do you address this issue and rebuild momentum? The first step is to align your organizational mass and ensure your business velocity is accurately assessed.

Organizational mass represents your energy management; it's the stored energy within your organization, encompassing your vision, values, and structure. If this mass is misaligned, it leads to wasted energy, resulting in employee attrition, loss of time, and drained resources.

Business velocity requires you to reassess your current position and determine where you need to be to maintain balance and alignment with your organizational mass. Time plays a critical role in shaping your business velocity. Allocate too little time, and you risk prematurely propelling your organization forward without adequate preparation. Conversely, taking too much time can slow your business velocity and dampen momentum.

To effectively rebuild momentum, focus on the following:

1. **Realign Organizational Mass**: Revisit and reinforce your organizational vision and values. Ensure that everyone in your organization understands and is aligned with these core principles.
2. **Assess Business Velocity**: Conduct a thorough evaluation of your current status. Identify gaps between where you are now and where you want to be.
3. **Time Management**: Allocate time wisely. Implement a timeline that allows for necessary adjustments while ensuring you're not stalling progress.
4. **Engage Your Team**: Communicate openly with your team about the changes and the reasons behind them. Gather their feedback and insights to foster a sense of ownership and involvement.
5. **Monitor Progress**: Regularly check in on both organizational mass and business velocity. Make adjustments as needed to stay on track and maintain alignment.

By taking these steps, you can effectively manage the forces at play in your organization, ensuring that you not only regain momentum but also set the stage for sustainable growth and success.

Organizational Alignment Increases Acceleration

Once you've stabilized your business momentum, it's time to shift gears and accelerate. To achieve acceleration, you must ensure that your people forces are accurately calculated and strategically positioned to guide the organization in a forward direction. These people forces will play a crucial role in decision-making processes that drive results.

Here's how to set the stage for successful acceleration:

1. **Define Motivations**: Clearly identify what motivates your team members. Understand their individual goals and expectations, as this knowledge will be pivotal in aligning their efforts with the organization's objectives.
2. **Assess Personality Types**: Balance the motivations of your team with the personality types that are essential for moving the organization forward. This alignment will foster collaboration and enhance productivity.
3. **Align with Organizational Mass**: Ensure that the motivations and personality types are in harmony with your organizational mass, which includes your vision, values, and structure. This alignment is critical for maintaining energy and focus as you accelerate.
4. **Empower Decision-Making**: Encourage your team to take ownership of their roles in the acceleration process. Empower them to make decisions that align with the organization's goals, fostering a sense of accountability and commitment.
5. **Monitor and Adjust**: As you accelerate, keep a close eye on progress and be prepared to make adjustments as needed. Continuous feedback loops will help identify areas that require fine-tuning.

When you get this right, your organization will not only gain momentum but will also accelerate toward its goals smoothly and effectively. By harnessing the collective energy and talents of your team, you'll not only reach your desired destination but also create a journey that is both fulfilling and productive for everyone involved.

CHAPTER 25: Align Internal Mass for Momentum

To create a thriving organization, it's essential to establish a strong cultural foundation that aligns with both the organization's and the leader's vision, values, goals, and mission. This alignment of internal organizational mass is the start for building organizational momentum

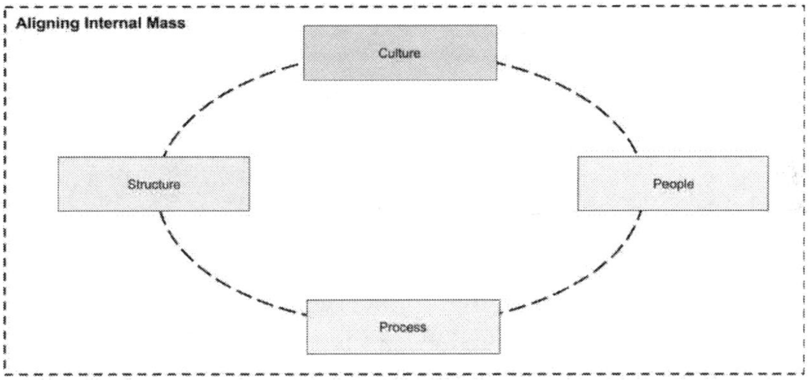

Figure 75. Organizational Culture alignment of internal mass.

The vision represents your organization's aspirations; it is the destination you aim to reach. Values act as the guardrails for the behaviors that guide your journey, ensuring that both you and the organization remain true to yourselves as you progress. Without these behavioral guardrails on your Road to Success, you risk veering off course, lacking the necessary safety measures to get back on track. Goals serve as the challenges that keep your organization anchored to your future vision, providing direction and purpose. Meanwhile, the mission outlines your target strategy, offering incremental measures to assess whether you're on the right path or if adjustments are needed. Together, your vision and values infuse your organization, its ideas, products, people, and processes with a clear sense of purpose.

Organizational Internal Mass

Internal mass refers to the organizational culture, encompassing your vision, values, goals, and mission. It reflects the acceptable behaviors within the organization, shaping how leaders and teams interact with one another and how they treat clients. Essentially, internal mass is the foundational framework that guides your organization's actions and decisions. The formula for calculating internal mass can be expressed as follows:

$$m \text{ (internal mass)} = \text{vision} + \text{values} + \text{goals} + \text{mission}$$

Figure 76. Internal Mass Equation

This equation highlights the importance of each component in creating a robust organizational culture that fosters alignment and purpose.

Embrace a Culture of Learning

To maintain an organization that thrives on innovation, it is essential to cultivate a culture of learning in everything you do. Implementing changes, adopting new strategies, and acquiring knowledge can feel overwhelming, especially when you're launching or revamping your business to foster progress. The rapid pace of change may leave you wondering how to ensure that your company's vision and values will endure over time.

Here's the good news: you can create a vision and a set of values, establish clear goals, and mobilize a mission grounded in sound strategy and decision-making, all of which can stand the test of time. As long as each component of your organization's culture is dedicated to continuous growth, learning, development, and education, you can achieve lasting success.

In the following sections, I'll share strategies for keeping your culture innovative at every level and for evolving it as new insights emerge. The key lies in ensuring that your processes align seamlessly with your internal mass, creating a dynamic environment where innovation can flourish.

How to Align Organizational Culture with Value Learning

As shown in the Four Levels of Competency Matrix (see **Figure 77**), when you begin learning something new, you start in the phase of "unconscious incompetence," where you are unaware of what you don't know. As you progress, you move to "conscious incompetence," realizing that you lack knowledge in a specific area, even if you're unsure about what that entails. Next comes "unconscious competence," where you develop the ability to perform the task without fully understanding the underlying principles. Finally, you reach "conscious competence," meaning you are aware of your actions, understand why they work, and can consistently replicate your success.

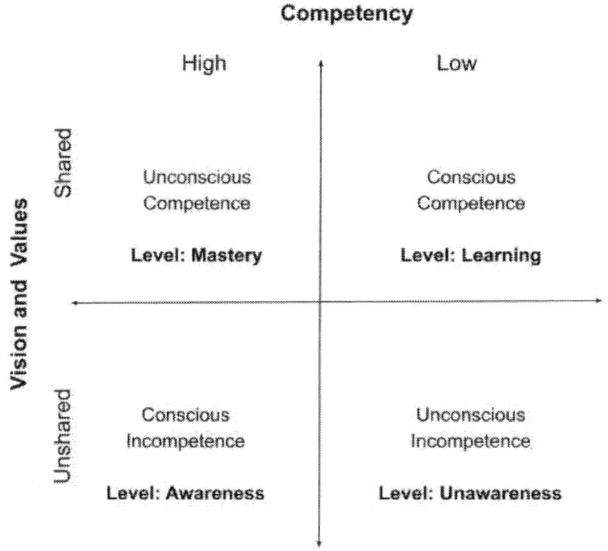

Figure 77. Four Levels of Competency Credited to Gordon Training 6International by its employee Noel Burch in the 1970s.

This mental model effectively captures the essence of learning within a forward-thinking environment. It can also serve as a powerful tool for recognizing and recruiting new talent, as it highlights the different stages individuals experience on their learning journey. By understanding where potential team members fall within this model, you can better support their growth and foster a culture of continuous development.

Levels of Learning and Competency Behaviors

When you start hiring individuals who align with your vision and values, you're tapping into the wisdom that propels your organization forward. This collective wisdom fosters momentum and growth. The journey toward wisdom encompasses four stages of learning, and your goal is to identify candidates who resonate with the foundational beliefs and guidelines you've established. Think of business as a game; learning how candidates think and approach challenges should be treated as part of that game. Their thought processes will enable new employees to learn and contribute effectively to your organization.

The journey begins with unawareness and progresses toward mastery. Awareness acts as the catalyst for growth. Your aim is to help individuals transition from "unconsciously incompetent" behaviors to "consciously incompetent," where genuine learning begins. With practice, they advance to the "consciously competent" stage before ultimately reaching "unconscious competence." This learning framework illustrates how mindset shifts are central to Mindset Mechanics, highlighting that awareness, practice, and perseverance are essential for growth in leadership and other vital skills.

Unawareness Level is Unconscious Incompetence

Unawareness represents the absence of knowledge and truth, often mistaken for ignorance—a term that carries a negative connotation because it typically targets the individual rather than their behavior. The reality is that we all have gaps in our knowledge, fluctuating between states of unconscious incompetence. Some individuals remain in this state, while others recognize their ignorance and choose to change based on how important that change is to them.

Behavior associated with unconscious incompetence can be likened to playing the role of the fool. We all find ourselves in this position from time to time, whether loudly or quietly. However, it's vital not to linger in this role for too long. Unconscious intelligence underscores the importance of training, education, and continuous learning within your organization. When hiring new employees, it's crucial to quickly assess whether they are willing, unwilling, or unable to learn, grow, and contribute effectively. Candidates who are unaware of their need for growth can hinder your organization's progress. Thus, screening potential employees is essential to avoid wasting time and resources on interviews that won't yield the desired results.

Individuals exhibiting unconscious incompetence often fail to recognize the gaps in their knowledge or capabilities. Those unaware that their actions do not align with your organization's vision and values pose a significant risk. Their inability to see the value in acquiring new skills or altering behaviors leads to mistakes and ineffective actions, which they may not recognize due to their lack of awareness. This lack of insight can result in overconfidence; they may harbor false confidence in their abilities because they don't recognize their limitations. For instance, a new manager who has never received feedback on their communication style may believe they are clear and motivational. However, the team may perceive them as confusing or unapproachable, and without awareness of this disconnect, the manager remains in a state of unconscious incompetence.

I personally strive to approach situations with the mindset of knowing nothing and remaining open to continuous learning. Accepting new information is a process, especially since we all carry biases and preconceived notions that aim to protect us. I recall being advised at twelve years old to be cautious about what I read, as too much information could be harmful. In today's digital age, the idea that information poses a threat to society seems diabolical. When has information ever harmed anyone? I'm not referring to cases where witnesses are silenced to suppress crucial information but rather to the idea that "guns don't kill people; people kill people." The same principle applies to information; it doesn't cause harm—too little information does.

To transition out of the state of unconscious incompetence, individuals must embrace feedback or new information, which helps illuminate skill gaps. Exposure to different perspectives can spark awareness, as seeing others excel in the same skill area can serve as inspiration and provide role models. Additionally, practicing self-reflection is essential, as personal evaluation can help reveal blind spots in abilities or knowledge, guiding individuals in determining their readiness for the next level of growth.

Awareness Level is Conscious Incompetence

Conscious incompetence marks a pivotal stage in personal and professional growth, where individuals become aware of their limitations in specific areas, skills, or behaviors. This newfound awareness ignites a desire to learn and develop the necessary skills for improvement. Recognizing one's shortcomings is a profound step toward personal evolution, enabling individuals to better serve themselves and others.

When a person acknowledges their performance isn't up to par, it's a transformative moment. The real magic happens when they actively seek opportunities for growth, cultivating a routine that prioritizes self-improvement. While many feel frustration or discomfort in confronting their weaknesses, it's essential to embrace this unease. The journey of self-discovery, though challenging, can lead to exciting and rewarding changes. Focus on honing the skills that truly matter, and remain open to feedback and learning.

Consider the world of acting—a field where feedback is invaluable. Good actors thrive on constructive criticism, understanding that it fuels their growth. Whether positive or negative, feedback is a tool for refinement. For instance, an actor might receive comments about lacking emotional depth in a scene. Although it stings at first, this insight drives them to explore their character more profoundly, ultimately enriching their performance. Constructive criticism becomes a catalyst for breakthroughs, pushing them to dig deeper and enhance their craft.

Similarly, organizational leaders must adopt this perspective of awareness for themselves, their teams, clients, and even family members. For example, if a manager learns they struggle with delegation, they realize that micromanaging hampers both their team's growth and their effectiveness. They become consciously incompetent, aware of their limitations in delegation but not yet equipped with the skills to improve. This stage necessitates effort and commitment to address the issue or risk undermining their relationships and team morale.

To progress beyond conscious incompetence, individuals require structured learning environments and opportunities. Formal training, mentorship, and guided practice are essential for developing the necessary skills. Consistent feedback is crucial, as it helps correct mistakes and fosters better techniques. Moreover, a willingness to invest time and effort is paramount. Mastery comes from repetition, and this stage involves trial, error, and gradual improvement. Embracing conscious incompetence is a vital step in laying the groundwork for meaningful growth and transformation. It's an uncomfortable yet essential phase on the path to achieving competence.

Learning Level is Conscious Competence

Learning is a journey towards proficiency, and at the level of conscious competence, individuals begin to master skills while remaining acutely aware of the steps required for success. This phase is characterized by what you might call effortful execution. Although they can achieve desired outcomes, it necessitates deliberate focus, practice, and concentration. As they gain confidence, they realize the importance of maintaining attention to detail and a methodical approach to ensure accuracy.

At this stage, individuals can demonstrate their skills reliably, meaning they can execute tasks well but must consciously think through each step to maintain their performance. This conscious engagement fosters a sense of progress and achievement as they become aware of their growing capabilities and can see tangible results. Work transforms into a rewarding experience as they engage in deliberate practice, applying what they learn while utilizing feedback, reflection, and repetition to refine and reinforce their skills, ultimately improving efficiency over time.

Those who achieve unconscious competence are closely aligned with the vision and values of your organization. They are eager to learn and can complete tasks effectively, yet they still need time and attention to fully develop their decision-making skills in alignment with your organization's goals. Monitoring their behaviors and performance is crucial to prevent isolated decision-making, but continuous training and guidance are essential for enhancing their proficiency.

Consider a manager who has been diligently working on their delegation skills. Initially, they struggled (conscious incompetence), but through coaching and practice, they have developed the capability to delegate tasks effectively. Now, when they delegate, they carefully plan, choose the right team members, and ensure clear communication. While they can perform these tasks well, they still need to take the time to think through each step—they are consciously competent in delegation.

The path to unconscious competence is achieved by reviewing information and repeatedly performing the skill, allowing it to become second nature. By applying the skill across various contexts and scenarios, individuals deepen their understanding, making execution more natural and automatic. With increased familiarity comes confidence, leading to automatic high performance.

Mastery Level is Unconscious Competence

Mastery is the ultimate goal in any learning journey. When an individual achieves skill mastery, they execute tasks instinctively, without actively thinking through the steps involved; their behavior becomes seamless and fluid. This is the desired state for all employees aligned with your organization's vision, values, goals, and mission. Actions become natural, consistent, and often performed with ease, requiring little conscious effort because the skill has been practiced so frequently that it is ingrained, allowing for accurate performance without deliberate focus.

To reach this level of intuitive performance, one must act quickly and effectively, establishing behaviors that become part of their muscle memory or mental patterns. Mastery is evident when individuals can multitask effectively, as the primary skill no longer requires concentrated attention; they can perform it while managing other tasks. This level of mastery instills high confidence, as proficiency and consistency in the skill yield desired results that others also recognize.

As a leader, your goal is to guide each employee toward mastery in their respective roles. At this level, individuals feel valued and significantly contribute to the organization's overall performance. I believe all organizational leaders have a responsibility to cultivate their own mastery and facilitate their teams' growth, not only for the betterment of the organization but as a social responsibility to improve the world we live in. Since we spend a substantial portion of our day at work, whether in businesses or nonprofits, it is our duty to help others become masters, leaving the world better than we found it. The organizational leader is the true hero—mentoring, guiding, and shaping the lives of others through interactions and coaching. The profound impact of these connections cannot be fully appreciated without recognizing the critical role organizational leaders play in enhancing the quality of our societies.

Consider a seasoned leader who has mastered public speaking over many years. Initially, they relied on notes, rehearsals, and prompts to deliver a speech. Now, they can engage an audience, read their responses, and adapt their message effortlessly. This leader is unconsciously competent in public speaking—they no longer need to consciously focus on delivery, tone, or body language; it all happens naturally.

Repetition is the mother of skill. Regular application of a skill ingrains it, making it automatic. Structured feedback loops and ongoing assessments are crucial during the learning process to achieve unconscious competence, refining the skill to the point where it becomes second nature. Learning requires adaptation across diverse contexts. Practicing the skill in various situations reinforces it, helping individuals respond to changing circumstances with ease and build efficient performance. Once cognitive energy is freed from performing the task, individuals can complete tasks more quickly and direct their focus elsewhere. Mastery enhances decision-making as well; leaders who are unconsciously competent in key skills can make quicker, more accurate decisions because they understand the mechanics involved, allowing their mindset to be primed for artistic mastery of their skills.

The mastery stage represents the ultimate integration of learning, where the skill becomes part of an individual's core behavior and mindset. It is the ideal outcome for mastering any behavior or skill. Instead of playing the role of the fool, the leading role of the master is now achieved.

Learning Skills is Synonymous with Behaviors

Learning is synonymous with behavior because it's reflected in how people act, respond, and adapt over time. A common belief is that learning is primarily focused on acquiring knowledge, which is partially true. When someone learns a new skill, it's not just about acquiring knowledge; it's about changing their behavior to incorporate that knowledge. Behaviors are the result of a mental state of being.

In other words, learning is the ultimate behavior that is demonstrated by what you do. Some individuals view learning about something as merely amassing a large amount of information. However, in truth, the **Learning-Doing Matrix** (**Figure 78**) demonstrates how learning also involves knowing how to put knowledge into practice before actually doing something. The only way to determine if you or your team has developed a new skill is by observing actions. What you do determines if learning has taken place.

Learn		Do
About something	*How to do something*	*Something*

Figure 78. Learning-Doing Matrix

For example, consider someone learning to play the piano. At first, their fingers move clumsily over the keys because they're unfamiliar with the notes and technique. As they practice, their behavior changes—they press the keys more confidently, their hand movements become fluid, and they can play entire songs without hesitation.

The shift from awkwardness to mastery demonstrates that true learning is evident and the behavioral transformation has occurred. It's not enough to know the notes; it's about how well you can execute them into action. This is why learning is so deeply linked to behavior—it's through changing what you do that learning becomes real and visible.

CHAPTER 26. Align External Structure for Momentum

To create a thriving organization, aligning your internal mass is just the beginning. The next critical step is building external mass, which encompasses your organizational structure and defines how your organization operates. This structure emphasizes function and design, outlining leadership roles and key functional areas.

As we discussed in Chapter 9, one of the foundational principles of architecture is that *form follows function*. This principle applies universally, from the atomic level to the vastness of the solar system. The shape of your organization directly influences its function, and conversely, the function often dictates its design. Often, when an organization isn't performing as desired, it can be traced back to a design flaw. In other words, the structure of your organization may not align with your goals and objectives.

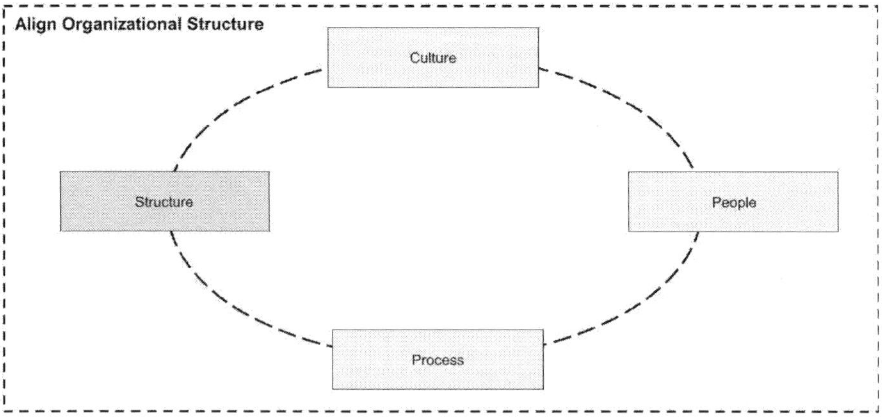

Figure 79. Align Organizational Structure

By examining and refining your external mass, you can create a solid foundation that supports your internal values, fostering a dynamic and effective work environment.

Organizational External Mass

External mass represents the organizational structure, encompassing both design and function. It defines your organization's purpose and how it interacts with its external environment. As a complex adaptive system, your organization must be agile, with a design that aligns with its strategic goals. In essence, external mass is determined by the equation:

m (external mass) = structure design + structure function

Figure 80. External Mass Equation

This means that how your organization is structured must support its purpose and effectively facilitate its operations. By ensuring that both elements work in harmony, you can create a robust framework that allows your organization to thrive in a dynamic environment.

Lex Sisney outlines organizational structure as the shaping and response of an organization to the environment by three criteria:

1. **Function:** the core areas and activities the organization must engage to accomplish its strategy (CEO, Operations, Engineer, Legal, Finance, etc.).
2. **Location:** where each function is placed within the organization and interacts with the other functions.
3. **Authority:** the ability of a function to make decisions within its domain of authority and perform activities without unnecessary friction.

How to Align External Mass

Understanding the importance of external mass is crucial for the success of any organization. External mass refers to the organizational structure, which encompasses both design and function. It defines how your organization operates and interacts with the external environment. This section explores common pitfalls organizations encounter when implementing change and the strategies for overcoming inertia to foster growth and innovation.

External mass consists of organizational design and function. It encapsulates how your organization exists to serve its purpose and interact with its surroundings. Your organization is a complex adaptive system that must be agile, with a purpose (design) that supports its strategy (function). The formula for external mass can be defined as external mass equals structural design plus structural function.

A common issue is that organizations become stuck in their ways, causing inertia to build up. New initiatives and missions are introduced, but if the structure remains unchanged, it stagnates growth, deflates momentum, and increases frustration. This is akin to putting "new wine into old wineskins" and expecting the organization to grow without bursting.

Here are some common mistakes to avoid:

1. **Not Adjusting Structure**: Whenever your strategy changes or shifts to a new execution lifecycle, it's crucial to reevaluate and adjust the structure. Change occurs from the top down. Restructuring can help alleviate resistance to structural change and enable everyone to see the bigger picture, thus understanding and appreciating their new roles.
2. **Narrow Focus on Effectiveness**: Avoid overly fixating on executing tasks perfectly. Perfectionism can stifle spontaneity, taking the fun out of business and diminishing team vitality. Effectiveness requires space and time for creativity, so structuring roles prioritizing efficiency can hinder innovation.
3. **Short-Range Results Blocking Long-Range Development**: Organizational structure should follow a top-down approach while being driven from the bottom up. Effectiveness should drive efficiency, strategy should drive change, and long-range development should influence short-range results. For instance, you wouldn't want Product Development (a long-range function) to report to Operations (a short-range function), as this would force the product development team to focus on immediate operational needs rather than long-term customer requirements.
4. **Imbalance of Autonomy and Control**: There's often a natural tug-

of-war between autonomy and control. Functions closest to the customer (like sales, customer service, and marketing) require more autonomy, while those closer to the product or service (like operations, product development, and branding) should have more control. Functions focusing on systems (like accounting, legal, and HR) should balance autonomy and control to cultivate each function's strengths for the organization's benefit.

5. **Placing the Wrong People in Functional Roles**: Your organization's performance is similar to a system; the output is only as good as the input. Many organizations make the mistake of placing warm bodies in open seats without considering whether these individuals are a cultural or functional fit. The personality types in each role must match the required organizational forces. As in sports, you want the best players in the right positions to maximize value to the team and organization. Sometimes, simplifying operations and waiting for the right person is the best strategy for developing your organizational structure.

To illustrate how an effective organizational structure can benefit your business, we can look at Apple's organizational structure as an insightful example.

Apple's Secret Sauce

Apple's innovation success didn't appear from thin air. Its success is largely due to how it's organized internally. Before Steve Jobs' return in 1997, Apple's organizational structure was traditionally organized by business units with separate profit & loss (P&L) responsibilities. This stifled innovation. Jobs decided to organize his team by function, with a single P&L responsibility aligning expertise with decision-making rights, as seen in Apple's organizational chart when Steve Jobs was CEO (**Figure 80)**. To this day, Apple is still organized this way. But why is Apple structured this way, and how does it help with innovation capabilities?

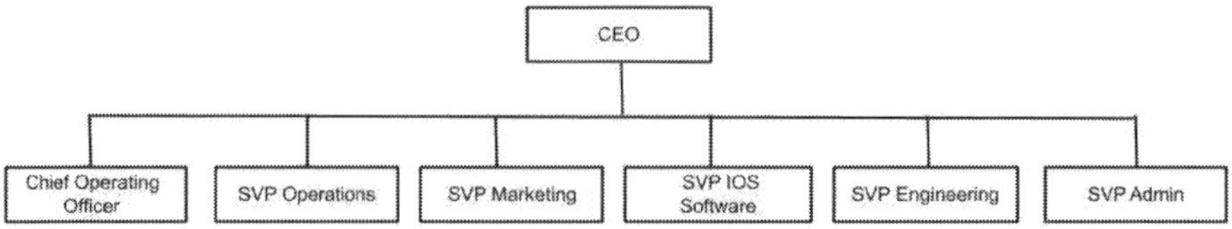

Figure 81. Apple Organizational Structure of SVP Level

Apple is organized like a lean startup. At the core of the organization's values, Apple believes that design rights should remain with the people who are the most knowledgeable and experienced experts in their domains. Each functional leader is an entrepreneur without the risk. Another reason why Apple is structured this way is because expertise aligns very well with functions; the same thing cannot be stated, which isn't the case for products or divisional structures. This allows Apple's senior vice presidents, not general managers, to be in charge of functions, not products.

Apple's structure is based on two views.

1. Given the fact that Apple competes in highly dynamic and disruptive markets, to remain competitive, it's necessary to make frequent technological and design changes before products hit the market. The company must rely on the judgment and intuition of technical experts who can predict which technologies and designs are likely to succeed. General managers are most unlikely to have the capabilities to do this.

2. Apple could not do what it does if costs and price targets were fixed parameters for design and engineering choices. Instead, R&D leaders are encouraged to weigh the benefits of users against cost considerations.

To understand the power of effective organizational structure, let's examine Apple's approach, particularly in innovation and decision-making. A notable example is the introduction of the dual-lens camera in the iPhone 7 Plus in 2016. Paul Hubel, a senior leader involved in this project, and his team took significant risks with this feature. If users were unwilling to pay a premium for a phone with a higher-quality, more expensive camera, it could undermine the team's credibility for future proposals involving costly upgrades or features. However, the dual-lens camera became a defining characteristic of the iPhone 7 Plus.

Under a traditional organizational structure, Hubel may not have been empowered to take such a bold risk. The feature likely wouldn't have been developed due to concerns about costs, price analysis, and a lack of deep insight into user needs. When leaders making decisions have profound expertise in their fields, they can better balance costs and the value added to the user experience. This shift in decision-making underscores Steve Jobs' vision to change how Apple operates, emphasizing that the best individuals to make such impactful changes are specialists rather than general managers.

To implement this structure effectively, Apple requires its managers to possess deep expertise in their functional areas, encouraging them to dive as deeply into the details as necessary. The company comprises hundreds of specialized teams collaborating and making collective decisions for the greater good. Tim Cook, Apple's CEO, recognizes that "collaboration and joint decision-making are essential components of Apple's innovative management approach." This strategy enables Apple to engage resources to foster alignment between expertise and decision-making rights, contrasting with conventional organizational structures that often prioritize strict accountability and control.

Apple's organizational and leadership model saved the company from bankruptcy and propelled it to become one of the world's most influential technology firms. It is an excellent model for those looking to start and grow an innovative organization capable of swiftly and effectively implementing change.

Simple Functional Structure

Description. The functional form divides work by type, e.g., marketing, finance, production, and administration. Although the functions may vary from industry to industry, the structure's organizing concepts are the skills needed to perform clusters of tasks; plants are assigned to the manufacturing function, sales perhaps in its own "arm," and marketing perhaps as a separate division. The underlying assumption in a functional structure is that the key strategic organizational capabilities lie in various functional skills—and hence, they are emphasized and central. Functional organizations tend to be centralized in that it is only at the senior level that the melding of the activities of the various functions occurs, hence the term "general management." A simple, functional organizational chart appears in **Figure 82**.

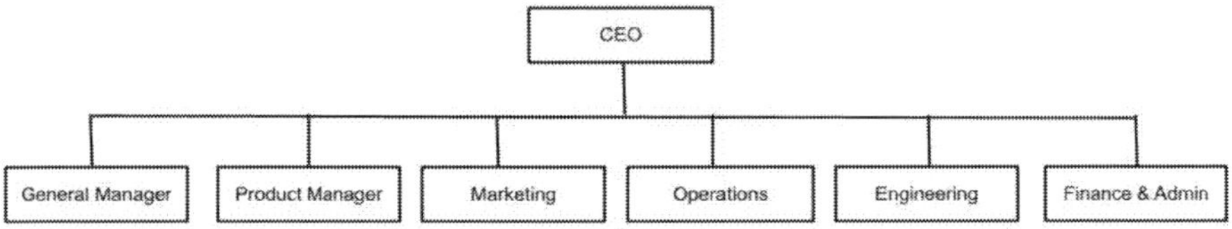

Figure 82. Functional Structure Example for Lean Organizations

Advantages. The functional form's primary advantage is the expertise of its functional specialists. People who begin their careers in research and development, manufacturing, or sales and work their way up through career ladders in those functions become practiced and skilled in their area of competence. Management's job is to integrate the functions and ensure they are coordinated in delivering their goods and services to the market. Hiring and retaining functional specialists is also a critical need. Functional forms make it easier for people within and without the organization to know whom to call for particular needs. They avoid duplication of resources and permit economies of scale in using the various specialties. Finally, the functional organization is simple—each department has clear assignments so that accountability is easier to track. Many smaller organizations begin with functional organizations and then evolve into one or more of the other types.

Disadvantage. On the other hand, the functional form often tends to encourage empire-building and protecting behavior in that functional managers often find it difficult to relate to or support the objectives of their counterparts. Boundaries between departments in a functional organization become increasingly difficult to penetrate. Customers and employees may hear people saying, "I'm sorry, that's not my department"—what well-known speaker and author Ken Blanchard calls "bureaucratic quackery." It can be easy to lose sight of customers in a functional organization—employees tend to be focused on their "jobs" rather than their customers. The functional form also creates barriers to coordination and communication among the functional units as units develop their own jargon and reward systems. Each functional group tends to develop its own subculture as well. A major problem is finding integrative systems and mechanisms that are sufficiently strong to overcome functional barriers can be a major problem. Finally, functional forms are not conducive to the early development of general management talent. To overcome these disadvantages, the CEO must remain aware of the overall culture by ensuring the organizational vision is aligned and integrated into the daily work behaviors of the entire team.

Alternative Structure

As your organization grows and increases its product line offering, the next organizational structure that allows for the most innovation is the functional-product hybrid form.

Description. We can readily identify various hybrid types of organizational structures in society. Many firms mix the features of two or more structures to maximize benefits. Obviously, the danger is when an organization's structure is built on disadvantages instead of strengths. Hybrids often build on one of the two forms by adding various mechanisms to increase integration—like task functions and work teams, which are frequently used to coordinate activities among divisions of a functional or product-form organization.

The most common hybrid forms include (a) the functional/product, where product managers don't have complete control over the operations or financing parts of the business, and (b) the geographic/customer type, where the area managers also have customer segment breakdowns in their sales forces (see **Figure 83).**

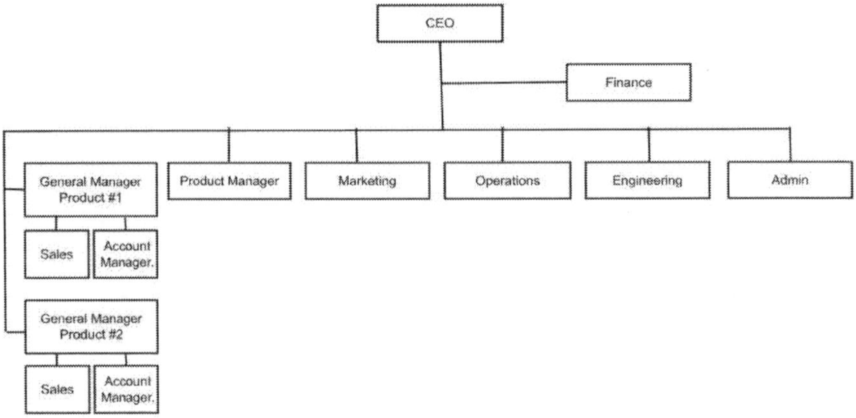

Figure 83. Functional-Product Hybrid Structure Example for Lean Organizations.

Advantages. Hybrids provide the means for adapting an organization's structure to the demands of the environment, current staffing characteristics, and other design constraints like in-place systems and historical momentum. The options are almost limitless, yet most organizations tend to revolve around one or more of the seven previously outlined.

Disadvantages. Hybrids can be confusing. Managers and employees, people in general actually, usually prefer to reduce uncertainty, and the more complex an organization is, the more ambiguous is the experience of each individual manager. In that confusion about *"who's doing what and reporting to whom about what"* people working in hybrids can become frustrated and disillusioned. Keeping the organizational structure simple aids in lessening ambiguity.

Conclusion

Organizational structure is a critical element in ensuring the effectiveness of any organization. It serves as a common lever that many executives pull in an effort to enhance efficiency. However, if an organization's structure does not align well with its environment and internal systems, it will struggle to operate at high levels of effectiveness and efficiency. Unfortunately, a universally appropriate organizational structure for any situation is seldom available because the structure is about design and purpose. Managers tasked with redesigning or influencing organizational design should consider essential factors such as fit, differentiation, integration, technology, size, span of control, and line authority while crafting structures that work.

In this evolving landscape, leaders face a diverse array of alternatives and a vast arena of possible variations. The rise of new Information Age organizations has brought about a shift where powerful information networks distribute greater decision-making power across all levels. These "infocracies" empower individuals who have access to customers and information databases, enabling them to merge insights and make sound business decisions.

The evolution of these new organizations will be driven by the creativity of their leaders and managers and their innovative approaches to organizational design. Gareth Morgan, in *Images of Organizations,* offers intriguing metaphors for thinking about organizations, suggesting they can be perceived as machines, living organisms, brains, cultures, political systems, psychic prisons, foci of constant change, or instruments of dominance. Our perspectives on organizations undoubtedly shape our design decisions.

In *Organizing for the Future,* Jay Galbraith and Ed Lawler compile insights on creating highly engaged organizations that unleash the collective creativity of employees. Such ideas resonate with managers concerned about the potential bureaucratic entrapment within their corporations, especially amid the surge of e-commerce companies emerging today. Peter Drucker has posited that future work will revolve around projects, a view echoed by Tom Peters in *The Project 50,* which outlines the traits of effective project managers. This model advocates for a decentralized design, as seen in companies like H-P and Johnson & Johnson, keeping business units small enough to foster creativity while maintaining high levels of coordination.

This overview introduces you to the significant organizing principles developed over the past two centuries, the types of organizational structures commonly employed in the 20th century, and the importance of structure as a pivotal decision point for leaders striving to build effective organizations. It also encourages you to think about organizational hybrids—blending historical insights on design and theory as we enter a new, information-driven millennium. As organizational forms continue to evolve, it's vital for astute leaders to stay informed about these developments and leverage them to optimize employee efforts toward achieving strategic objectives.

---

49. Bretz, Lukas. "Principles of Organization Structure." Arabian J Bus Manag Review 12 (2022):442. https://www.hilarispublisher.com/open-access/principles-of-organization-structure.pdf
50. https://ils.unc.edu/daniel/405/Montana11.pdf
51. Pitts, Tammy, and James Clawson. "Organizational Structure." *University of Virginia Darden School Foundation*, vol. 10, no. OB, 2008, p. 19. *University of Virginia Darden School Foundation*, https://www.researchgate.net/publication/228144159_Organizational_Structure.
52. Reddy, Thanu S. "Apple's Secret Sauce: The Role of Organizational Structure in Its Success." *YouTube*, 19 June 2023, https://www.youtube.com/watch?v=4Vl4WHRhz9U. Accessed 7 October 2024.
53. "How Apple Is Organized for Innovation: The Functional Organization." *YouTube*, 27 January 2021, https://www.youtube.com/watch?v=5hENFA3CJUY. Accessed 7 October 2024.
54. Podolny, Joel M., and Morten T. Hensen. "How Apple Is Organized for Innovation." *Harvard Business Review*, November 2020, https://hbr.org/2020/11/how-apple-is-organized-for-innovation. Accessed 7 October 2024.
55. Limeri, Lisa, et al. "Chapter 18: Form Follows Function – Introductory Biology 2." *Raider Digital Publishing*, https://raider.pressbooks.pub/biology2/chapter/18-form-follows-function/. Accessed 7 October 2024.

CHAPTER 27. Align Business Velocity for Momentum

President John F. Kennedy's moon speech in 1962 galvanized the spirit of innovation in the United States and set the course for a monumental achievement in human history. He famously declared that the nation would reach the moon "not because they are easy, but because they are hard" and "because it is there." Though he did not live to see the event, his words ignited a fire of ambition in countless individuals who contributed to the successful moon landing on July 20th, 1969. This chapter explores the power of inspiring your teams to fulfill the organization's vision, values, goals, and mission. The final component crucial for achieving organizational momentum is aligning *Business Velocity*.

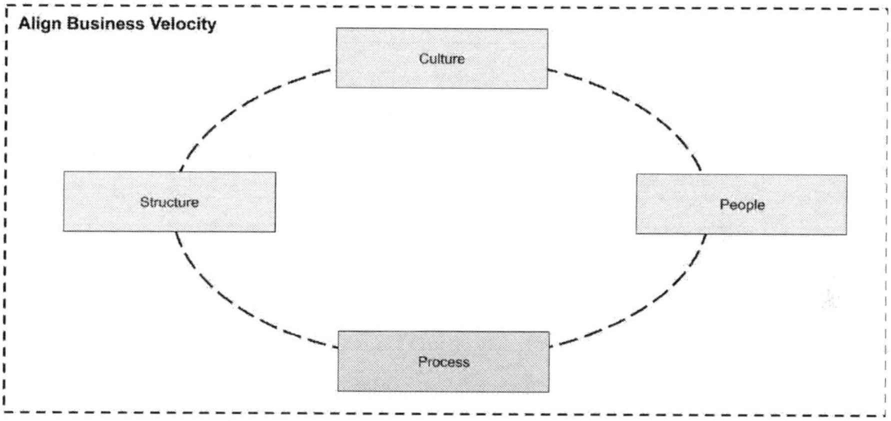

Figure 84. Business Velocity momentum consists of decision-making and implementation processes.

As discussed, business momentum can be defined by the formula mass times velocity. Now, let's focus on understanding the formula for business velocity itself. Business velocity represents the aligned decision-making process within the organization. How decisions are made can either accelerate or impede results. However, it's not merely about making quick decisions; what truly matters is the quality of those decisions. Organizational leaders must take the time to make the right choices and then move swiftly to implement them. Therefore, business velocity consists of a two-step process: deciding and implementing.

Business Velocity

Business velocity is defined as the change in displacement over time. This formula emphasizes the importance of measuring how quickly and effectively an organization can move from its current state to its desired state. By understanding this concept, leaders can better assess the pace of their decision-making processes and implementation strategies, ensuring that they are moving quickly and in the right direction. As shown in the equation:

$$v \text{ (business velocity)} = \left(\frac{\Delta x \text{ (displacement)}}{\Delta t \text{ (time)}} \right)$$

Figure 85. Business Velocity is displacement over time.

Displacement represents the transformation process within an organization, defined as the difference between your current location and your desired location. This concept captures the essence of change and growth, highlighting the journey from where you are now—facing challenges such as overwhelm, team conflicts, and missed opportunities—to where you aspire to be, characterized by vitality, high performance, and sustainable growth. Understanding displacement is crucial for leaders, as it helps them identify the gaps that need to be bridged to achieve their goals and drive their organizations forward. As shown in the equation:

$$\Delta x \text{ (displacement)} = \text{desired location} - \text{current location}$$

Figure 86. Business Displacement Equation

Your current location represents your being overwhelmed, having team conflicts, missing opportunities, and facing various challenges. In contrast, your desired location is where you aspire to be—a place filled with vitality, future opportunities, a high-performance team, and sustained growth. Time is critical, facilitating decision-making and implementation as you journey from your current location to your desired destination. It works in your favor, impacting the space-time continuum.

In this context, decision-making and implementation are essential seeds for progress. They are integral to the seedtime and harvest process, following the rhythmic cadence of *go slow to go fast.* This approach emphasizes the importance of thoughtful planning and execution, allowing you to cultivate the right environment for growth and success.

The Cadence of Decision-Making and Implementation

From the outset, achieving speed in execution hinges on a harmonious balance between effective decision-making and timely implementation. However, there's an important caveat: these processes must occur at the appropriate organizational levels. If the CEO is responsible for every decision, they risk becoming a bottleneck, inadvertently stifling innovation and responsiveness. When this happens, the CEO shifts from a vital leader—who drives innovation, provides oversight and automates decisions through processes and people—to a functional leader.

A vital leader empowers their team, fostering an environment where team members feel confident in making decisions. If team members constantly seek direction from the CEO, they may hesitate to take the initiative, leading to diminished confidence in the leadership's authority. This dynamic can cause operations teams to circumvent their immediate leaders, ultimately diverting valuable time and attention from the CEO.

As an organizational leader, building your structure with the end goal in mind is crucial. Whether starting up or expanding your business, you must embody behaviors and intentions aligning with your desired destination, not just your current circumstances. It's essential to streamline your processes so that decision-making becomes seamless and intrinsic to your operations, enhancing your ability to serve customers and foster collaboration within the team.

Your role as CEO involves ensuring that decisions and actions are managed at the appropriate levels by the right people, all while adhering to processes that fit your organizational culture. To achieve this, you must communicate the levels of authority, power, and influence assigned to each functional leadership role, positioning yourself as the vital leader who empowers others to thrive.

Getting the Right People in the Room

Knowing who to include in decision-making ensures effective decisions and swift implementation. To accelerate progress, it's essential to identify individuals who will be impacted by a decision and involve them early on in the process. Engaging diverse perspectives enriches the decision-making experience and enhances the quality of outcomes.

For optimal acceleration, you should incorporate the organizational forces represented by four key roles, each bringing a unique viewpoint to the table:

1. **Guardian Implementor**: This individual understands what needs to be done to execute the decision effectively. Their focus on practical execution ensures that strategies are actionable and grounded in reality.
2. **Nurturing Unifier**: This person empathizes with the decision's impact on others. They help maintain morale and foster collaboration, ensuring team dynamics are considered and supported throughout the process.
3. **Connecting Communicator**: Equipped with valuable information about implementing the decision, this individual excels at bridging communication gaps and facilitating understanding among team members. They ensure everyone is aligned and informed.
4. **Creative Innovator**: This role is critical for identifying alternative paths to achieve the same results. The Creative Innovator challenges conventional thinking and explores innovative solutions, which can lead to more effective and efficient outcomes.

By assembling a team that includes these diverse perspectives, you'll create a more comprehensive decision-making process that accelerates implementation and enhances the overall effectiveness of your organizational strategies.

Then, you must ensure those chosen align with the organizational culture—internal and external mass. This can be seen through their domain of authority, amassed power, and influence.

1. **Authority**: Authority is the legal right to decide. This is granted by the organizational leader, or CEO, to a functional role.
2. **Power:** Power is the ability to help or hinder. This is earned by their ability to implement with skill mastery, effectiveness, and consistency.
3. **Influence:** Influence is the ability to get what you want without relying on authority or power. This is given by the team to individuals who have proven themselves to be reliable

What Does an Effective Business Velocity Look Like?

In the movie *Men in Black*, there's a memorable scene where Agent Jay, played by Will Smith, finds himself in a room filled with some of the brightest minds, all gathered to join a covert government agency. As they fumble with their paperwork, trying to write in the palms of their hands, on their laps, and even in the air, it's clear that many in the room haven't given much thought to the next steps. In contrast, Jay exemplifies effective decision-making. He engages with the right people, asks insightful questions, demonstrates thought leadership, and ultimately takes action, embracing responsibility while encouraging the group to reflect and integrate their ideas.

This brings us to the concept of "go slow to go fast," emphasizing the importance of carefully considering all decisions before committing to them. This approach requires thorough preparation, gathering as much relevant information as possible, and consulting those with deep knowledge of the situation. By collecting data and generating insights, you can make informed decisions followed by clear action steps assigned to key players for implementation.

The process doesn't end there; reinforcing, reflecting on, and integrating the decisions into action is crucial. Debriefing the outcomes allows you to assess whether the results align with expectations, providing valuable insights for future decision-making.

I recommend using the PATHS Method to navigate this effective business velocity—combining decision-making with implementation. This framework helps ensure that each step is deliberate and aligned with your organizational goals, facilitating a smoother and more successful execution.

PATHS Method to Business Velocity

The PATHS Method is a practical model for speed of decision-making and implementation. Outlined as follows:

P - Prioritize
A - Actualize
T - Theorize
H - Harmonize
S - Summarize

Prioritize

Setting the stage for effective decision-making begins with prioritizing the right people in the room—those who embody the necessary organizational mass and possess the right mix of personality forces. These individuals are often part of your leadership team and are crucial for driving meaningful change.

As you prioritize, it's essential to eliminate any inertia by clearly defining the purpose and destination of your discussions. This step is critical for securing full participation from those with the authority, power, and influence to implement organizational changes. Once you've successfully released inertia and engaged everyone involved, you can prioritize expectations and establish a timeline for future steps within the PATHS Method.

Leveraging the 80/20 Pareto Principle is key here. Aim to spend 80 percent of your time in the decision-making phases—prioritizing, actualizing, and theorizing—while reserving the remaining 20 percent for the implementation phases—harmonizing and summarizing. This structured approach ensures that the groundwork is solid before moving forward, setting your organization up for success as you navigate the complexities of change and innovation.

Actualize

The next step is to actualize the decision-making process. This involves gathering data and generating insights by asking challenging questions that reach the core of the issue, challenge, opportunity, or situation. The goal is to stimulate critical thinking among the participants and put them in the right mindset to provide valuable input based on their areas of expertise.

Once you have engaged everyone in the discussion, ensuring that all participants are aligned and on the same page regarding the current situation is crucial. Be sure to paint a vivid picture of the desired destination, igniting their enthusiasm to embark on this journey. Then, help your team see the benefits of reaching this destination, which will encourage them to participate actively in the process.

Additionally, it's important to clearly communicate the expectations for each participant upon arriving at the desired outcome. This clarity fosters buy-in and allows team members to offer insights on what might be missing, unnecessary, or needed for the benefit of all involved.

A common reason people struggle to buy into a vision or mission proposed by organizational leaders is not a lack of desire to contribute but rather confusion about the intended outcome. As I once heard, "A confused mind always says 'no.'" Your responsibility is to uncover the 'yes' that builds excitement, engagement, and participation throughout the process, ultimately accelerating progress towards your shared goals.

Theorize

Once the right people have a clear mental picture of the situation, it's time to theorize. This stage is marked by creating a framework for action. Effective decision-making leverages the collective thinking of everyone in the room to outline a series of steps for what should happen next.

Theorizing doesn't necessarily mean that everyone has reached the same conclusion; rather, it signifies that there is active participation and a consensus on the data presented, insights shared, and expertise available. This collaborative approach ensures that all perspectives are considered, fostering an environment where innovative solutions can emerge. By integrating diverse viewpoints, you can create a robust action plan that not only addresses the current challenges but also aligns with the organization's vision and goals.

Harmonize

Now that a basic theory has been established to frame the group's decision, the next step is to harmonize. In this phase, the group must decide how to collaborate effectively to test the validity of their theory. Given the insights shared, the participants present, and each individual's level of passion, tasks should be divided and assigned to ensure clear responsibility and accountability.

When delegating tasks, clarity is paramount. It's essential to set specific deadlines, outline expectations regarding authority, and address any potential conflicts of interest or scheduling issues that could arise later in the process. By fostering a collaborative environment where everyone understands their role and responsibilities, you can create a cohesive team ready to move forward with confidence.

Summarize

Summarizing is the crucial final step in the decision-making process. It ensures that nothing has been overlooked and that everyone is aligned with the mission at hand. This phase allows for the incorporation of any final objections, insights, or facts, ensuring that business velocity is optimized for success. By reviewing the key points and confirming understanding among all participants, you can create the best-case scenario for accelerating progress and building momentum.

Putting it All Together

Now it's time to build momentum by integrating all the elements you've gathered. You've collected the internal mass of organizational culture—comprising your vision, values, goals, and mission—and combined it with the external mass of organizational structure, which includes design and function.

Next, you've established business velocity, defined as displacement, or the distance between your desired location and your current location, over time (see **Figure 76**). You've accumulated internal and external mass, and now you can create lasting momentum.

Remember, momentum is calculated as mass multiplied by the change in velocity, or displacement over the change in time. By aligning these components, you're poised to propel your organization forward with clarity and purpose.

$$\text{p (momentum)} = \text{mass}\left(\frac{\Delta \text{x (displacement)}}{\Delta \text{t (time)}}\right)$$

Figure 87. Business Momentum Equation.

CHAPTER 28. Align People Forces for Acceleration

In any organization, success hinges on the ability to curate and develop the best in others. A prime example of this principle can be seen in President JFK's vision for the moon landing. His profound ability to speak to the inherent desires within people turned an audacious goal into a collective mission. While momentum is essential for getting things moving, acceleration is where the magic happens—it acts as a powerful multiplier. To achieve acceleration, it's crucial to gather the right people with the appropriate skills in the same room, as this combination serves as the catalyst for accelerated growth.

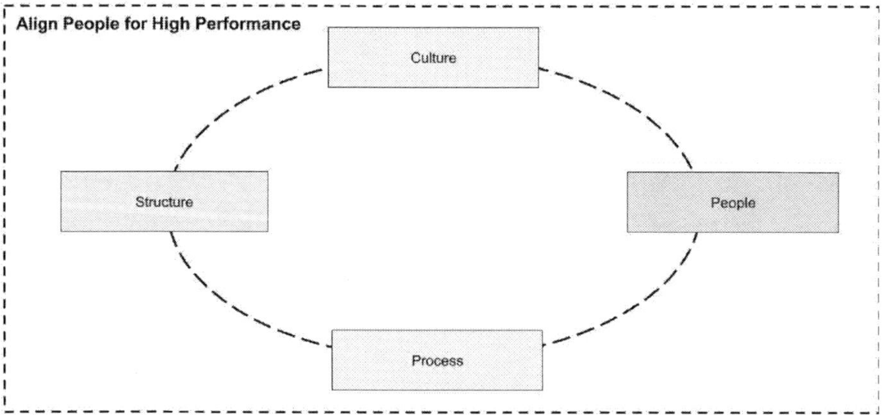

Figure 88. Align People for High Performance.

Understanding the desires of others and ensuring that those desires align with your organizational goals is key. Ultimately, it is the collective mass of these organizational forces that drives acceleration. Mathematically, acceleration can be represented as the sum of net (organizational) forces divided by the total (internal and external) mass involved.

$$\mathbf{a\ (acceleration)} = \left(\frac{\Sigma\ \mathbf{F\ (Force)}}{\Sigma\ \mathbf{m\ (mass)}} \right)$$

Figure 89. Business Acceleration Equation.

As a poignant illustration, consider the biblical scripture, "How could one pursue a thousand, and two chase away ten thousand?" (Deuteronomy 32:30 EHS). In a business context, this translates to the idea that one individual can effectively serve 1,000 clients, while two can handle 10,000. Here, two individuals do not merely double the results—they amplify them tenfold. Following this logic, one might infer that three individuals could potentially serve 100,000 clients. The key to this exponential multiplier effect lies in leveraging the right people and aligning their desires with the overarching vision of the organization before acceleration is applied.

Organizational Forces Decisions & Implementation

As a leader, your desire for authority, power, and influence must be met with a strategic approach that mitigates risks. Your mission must be to identify the right people—those who embody the appropriate behaviors, skills, and values—to support solid decision-making and swift implementation. To retain control while fostering growth, place these individuals in functional roles where they can evolve into vital leaders. By cultivating an environment that keeps them engaged and operating at their best, you empower them to contribute their fullest selves in service to customers and clients.

Energy and Vitality

Listening to your team is vital. Pay attention to what drains their energy versus what revitalizes them. It's this new energy that serves as a powerful motivator. As the organizational leader, your goal should be to attract the best ideas, customers, products, and services—those that add value, complement weaknesses and amplify the strengths of each team member. If there's a mismatch between what inspires and energizes your team, alignment issues will arise, leading to a loss of focus, inspiration, engagement, and mastery in their respective domains.

Everyone is a Vital Person of Influence

When you find the right people and put them in their proper function, everyone has the potential to excel in their own way. As you build your organization, ensure that no one is sidelined; every individual should step up as a leader. Each person is either a leader within their area of authority or a leader of their own actions. In either role, leadership involves taking responsibility for both successes and setbacks. When outcomes deviate from expectations, or when they align with your goals, your team must fully own the results. They should consistently strive for high performance and seek opportunities for improvement

Pay Your People

When it comes to fostering high-performance cultures, elements such as salary, bonuses, vacation days, extra time off, stock options, and promotions all play crucial roles in motivation. At the outset, you might find it challenging to offer competitive salaries, so consider starting with fractional or part-time roles. These positions can evolve into full-time opportunities as your organization grows. This approach is not only practical but also appropriate for your development stage.

As your team expands, it's essential to remain vigilant about your organizational culture and the behaviors that become established as norms. If you're not careful, compensation may become the sole influence in keeping your teams engaged, learning, and growing. To create an environment where your team can truly thrive, you must understand what matters to them and cultivate a culture that supports their needs and aspirations.

The Priorities of Business

During my recent trip to Southwest Michigan, I had the opportunity to visit the headquarters of the Edward Lowe Foundation. I was truly inspired by the passion and innovation demonstrated by the team running this nonprofit. They have wholeheartedly embraced the values instilled by the foundation's founders, Edward and Darlene Lowe. This experience reignited my appreciation for the purpose and priorities of businesses.

The essence of business lies in honoring our Creator by adhering to universal principles, as outlined in Matthew 6:33: to seek first the Kingdom of God. By aligning ourselves with these divine principles, we find that everything we desire will be added for our benefit. Conversely, straying from these principles leads to struggle and hardship. These principles echo the laws of physics, which provide us with the freedom to move, an abundance of resources, and a quality of energy and life. By prioritizing kingdom principles, we can navigate our lives and businesses with reduced friction, disorder, and risk. Both kingdom and physics principles guide us in how to treat ourselves and others, manage money and resources, engage with our teams and clients, and invest in our children.

I believe that the priorities of business are threefold.

1. **Leverage.** Leverage is a force multiplier; it increases results without using much effort. A good business should be designed to leverage the efforts of the leader, team, customers, and clients. However, many businesses are designed to just get the work done, no matter the strain on those most affected. This causes them to spend too much time and energy manually trying to make the businesses work the way they want them to. We only have a limited time on earth, so it's important to create businesses that benefit everyone involved so that everyone can achieve their goals and live a better life.
2. **Lifestyle.** Lifestyle is subjective in meaning and varies from person to person. To me, it is all about relaxation, exploring new places, and finding joy so that I'm energized and can share that positive energy with others. It also involves staying mentally sharp to create fresh ideas and develop innovative products and services to keep my business running smoothly. Additionally, it means dedicating time to my children and supporting them in achieving their goals.
3. **Legacy.** Legacy is the final destination and the most important reason why we start businesses. This means creating a foundation for the next generation to build upon, strengthening what already exists, and exploring new opportunities without having to start

from the beginning.

CHAPTER 29. Troubleshooting Guide

Let's take a moment to troubleshoot some of the most common issues that can arise within your business vehicle. Just as a car needs regular check-ups to ensure it runs smoothly, your organization requires careful evaluation to identify potential problems. This process begins with a Mindset Check, followed by a Mechanics Check, ensuring every aspect of your business is operating efficiently.

First, we conduct a Mindset Check, which involves a visual, eight-point preventative maintenance inspection designed to pinpoint possible leaks within the system. This initial assessment helps us identify areas that may need attention. Next, we perform a Mechanics Check, connecting the entire business vehicle to a systems diagnostic machine. This machine collects data, monitors performance, and assesses whether any organizational subsystems are at risk of failure. Together, these checks provide a comprehensive overview of your organization's health, allowing you to address issues before they escalate.

Mindset Check

Dr. Adizes identifies several "diseases" that organizations can suffer from, many of which stem from the interplay between organizational culture and leadership mindset. These issues include the leader trap, incomplete priorities, corporate amnesia, internal friction, liquid crisis, loss of innovation, and poor team performance. One significant concern is the organizational leader trap, a stage where the leader or founder, despite their best intentions, inadvertently hinders growth and success. At this juncture, the leader may become the very barrier to progress.

Leader's Trap

It's essential to consider whether your organization is at risk of falling into the leader's trap. Questions to ponder include: Does the leader fear losing control? Will they maintain a commitment to the team's success? Can they allow others to lead and inspire in their unique ways? Are they willing to share responsibilities and equity with management? Is there a succession plan in place? Can the organization continue to function smoothly in the leader's absence?

Dr. Adizes emphasizes that organizational leaders should consider bringing in external expertise during the critical transition between the seedling and aging tree lifecycle stages—not before or after. This phase is often the most challenging in the corporate lifecycle, as the changes required are fundamental, and the leader may be perceived as the obstacle to necessary growth. Recognizing and addressing these potential pitfalls is crucial for fostering a thriving and adaptable organization.

Competing Priorities

Incomplete priorities occur when an organization fails to evaluate opportunities properly, leading to the pursuit of projects that may not be beneficial. This behavior can result in the organization taking on various initiatives, some of which may hinder progress and overwhelm leaders and employees. To effectively address this issue, organizations must establish a robust vetting process that differentiates between ideal clients and those better suited for referral to other producers and service providers. Additionally, mastering the art of saying "no" is crucial to maintaining focus on the most impactful tasks. This integrated approach will help optimize productivity and ensure that efforts are directed toward the most rewarding endeavors.

It's essential to recognize that all business is not good business. This is where opportunity management and systematic decision-making can play a pivotal role in developing an effective opportunity strategy. By designing frameworks for identifying and analyzing your business and personal goals, strengths, weaknesses, opportunities, challenges, and risks, you can enhance your decision-making process. Incorporating systematic analysis into your approach increases awareness and deepens your understanding of the various factors at play in making informed decisions or crafting strategies to maximize your goals.

Deciding whether to accept an opportunity becomes less challenging with well-defined assessment and screening criteria, providing clarity and perspective that measure progress and align with your organization's true priorities.

Corporate Amnesia

Corporate Amnesia is when a company forgets why it was created and its purpose. Its values and visions are rarely considered, and it finds itself competing in a large pool of competitors rather than embracing its journey that distinguishes and creates a lane established on credibility and authority. Fundamentally, your purpose statement works against corporate amnesia. It should answer the question, *"What's a compelling business story?" "Why do we exist?" "What's important about what we do?"* Your purpose statement is the magnet that attracts investors, customers, and team members. A company cannot forget its identity because purpose forms the roots from which your company grows.

Internal Friction

Internal friction arises when a company expends excessive energy and effort on basic decisions and task completion. This friction creates obstacles that impede progress, making work more challenging and potentially frustrating for employees.

When crafting plans and determining courses of action, leaders must consider the impact on others' time. This involves, not only finding ways to conserve time for team members, but also acknowledging how their decisions may be imposed on others. Leaders should recognize the significant power and influence they wield; even a casual remark can prompt employees to invest time, effort, and resources into pursuits that may ultimately be unproductive.

Additionally, leaders need to be aware of their own areas of limited knowledge. Some may mistakenly believe that their past achievements equate to a complete understanding of their organization. However, ignorance of crucial details can lead to negative consequences. Adopting a philosophy of simplicity is vital—creating a list of obstacles that obstruct progress and cause frustration is a critical first step.

So, what actions will be taken to eliminate these barriers and foster a more efficient, productive work environment?

Liquid Crisis

A liquid crisis occurs when a company generates sales but struggles to achieve profitability. This phenomenon, often referred to as the silent killer of businesses, can have dire consequences. Even a successful business making substantial sales may face challenges in paying bills on time, as profits do not always equate to cash flow. Ineffective cash flow management can severely hinder a business's overall success.

Cash flow represents the ebb and flow of money within a business, distinct from profits, which reflect the earnings after expenses are covered. The significance of cash flow lies in its ability to indicate the funds readily available to a business at any moment.

A recent study by U.S. Bank revealed that cash flow problems are responsible for the failure of 82% of small businesses. This statistic highlights that even if a company appears profitable on paper, it can still face bankruptcy without sufficient cash to cover its obligations. Furthermore, 40% of startups fail because they run out of cash, underscoring the critical need for effective cash flow management from the outset. A promising startup can collapse quickly without sound financial oversight.

Businesses often struggle with cash flow due to delayed customer payments and high overhead costs, such as rent, utilities, and salaries. These challenges necessitate a proactive approach to cash flow management to ensure sustainability and growth.

Loss Of Innovation

The decline in innovation occurs when a company shifts its focus from fostering new and original ideas to relying on acquisitions or offering generic products and services for growth. This loss of innovation manifests in various ways, including short-sighted thinking, a lack of an innovative mindset, fear of failure, insufficient innovation structure and processes, a poor understanding of customer needs, low motivation among teams, and inadequate budgets.

Not Thinking Long-Term. Many companies emphasize immediate goals over long-term planning, often channeling significant resources into sales and marketing to achieve short-term targets. This strategy frequently leaves little funding for research and innovation, ultimately hindering sustainable growth and success.

Lack of Innovation Mindset. Innovation often stalls due to the prevailing mindset among employees within the organization. It is essential to cultivate a culture that encourages innovative thinking at all levels, starting from leadership and extending throughout the organization. Companies should prioritize creativity, foster the generation of new ideas, and empower employees to think beyond traditional boundaries. According to econometric research, firms that prioritize innovation experience faster growth and higher profits compared to those that do not, highlighting the importance of an innovative mindset and effective communication of organizational goals.

Fear of Failing. Employees may become apprehensive and reluctant to engage in new tasks or projects that push their limits due to a fear of failure. This fear often stems from the organization's failure to provide adequate support and resources for success.

Lack of Innovation Structure and Processes. Achieving tangible results in innovation is challenging without a robust framework and defined processes. Many organizations lack a proper methodology or structured approach to guide their innovation strategies.

Not Understanding Your Customer's Needs. Innovation efforts often falter when organizations do not thoroughly understand their customers' wants and needs. This disconnect can lead to new products or ideas that fail to address real customer problems or that customers simply do not want. To achieve revolutionary advancements, organizations must grasp the thoughts, emotions, and behaviors of their customers and how they interact with other businesses.

Lack of Team Motivation. Innovation is inherently a team effort and cannot thrive in isolation. Having a supportive and committed team that is fully invested in the innovation process is crucial. The success of any innovation initiative relies on empowering individuals and teams. Motivated teams generate innovative ideas, provide fresh perspectives, and achieve successful outcomes. It's vital for internal innovation teams to be enthusiastic and passionate about creating new products and ideas.

Lack of Budget. A significant barrier to innovation is often a lack of sufficient budget or investment in innovative initiatives. Without the necessary financial support for activities like team building, brainstorming, prototyping, and quality testing, innovation projects are likely to stall and ultimately fail. To ensure long-term success, organizations must develop a dedicated go-to-market strategy that identifies and addresses the pain points of their target audience, allowing innovation to flourish without setbacks.

Mediocre Team

When a team fails to perform at its full potential or meet the size of the opportunity, it's likely experiencing one or more common symptoms. These can include freeloaders, unclear goals, lack of transparency, lack of accountability, and fear of feedback. Identifying and addressing these issues is crucial for fostering a high-performing team.

Freeloaders. Teams burdened by freeloaders do not operate at peak efficiency. Research from Gallup indicates that approximately 20% of employees are disengaged from their work, which can significantly hinder the productivity of their colleagues. Even those who are not actively undermining their teammates contribute little value, leading to demoralization within the group. While addressing this issue may be uncomfortable, neglecting it will only exacerbate the problem over time.

Unclear Goals. Without clearly defined goals and objectives—or worse, with goals that constantly shift—the chances of achieving success diminish greatly. Talented individuals may seek opportunities elsewhere when they perceive a lack of direction, resulting in the loss of valuable team members. Furthermore, when employees are uncertain about the purpose of their work, disengagement and dissatisfaction can arise, creating a chaotic and unproductive environment. Utilizing a goal-setting framework can help establish clarity and motivate your team toward successful outcomes.

Lack of Transparency. A lack of discipline and transparency can lead to wasted time and confusion within the team. Endless meetings and calls may create a false sense of productivity while actual work remains incomplete. Implementing a culture of transparency can clarify goals and strategies, fostering greater accountability and better results.

Lack of Accountability. According to the Harvard Business Review, a staggering 50% of organizational leaders struggle to drive accountability within their teams. Many leaders are hindered by a fear of being perceived as the "bad guy" or a lack of discipline. Without accountability, mediocrity prevails, seriously affecting performance. Failing to hold underperforming employees accountable ultimately misleads them and can surprise them during performance reviews. The most successful teams consist of members who hold themselves and each other accountable. Engage your team in establishing a standard of accountability and commit to maintaining it consistently week after week.

Fear of Feedback. It's important to recognize that not providing feedback—whether positive or negative—can be more detrimental than giving negative feedback itself. While team members are often eager to receive feedback, managers may hesitate to provide it. Overcoming this discomfort and making feedback a regular practice is essential. Studies show that young professionals learn 70% of their skills on the job, making positive reinforcement crucial for development. For more advanced employees, constructive feedback is vital for further growth. Research suggests aiming for a feedback ratio of 5:1—five instances of positive feedback for every piece of constructive criticism.

By addressing these symptoms head-on, you can pave the way for improved team performance and ultimately foster a culture of high achievement within your organization.

Mechanics Check

Once the Mindset Check is complete, it's essential to pivot our focus toward conducting a thorough Mechanics Check of the organization. This crucial step is vital for ensuring that every aspect of the organization is meticulously evaluated, guaranteeing the seamless functioning and optimal performance of its mechanics. By dedicating sufficient time and attention to detail in this assessment, we can proactively identify any potential concerns or gaps within the organization's operations.

The Mechanics Check involves a comprehensive examination of all components, from systems and processes to team dynamics and resource allocation. This meticulous evaluation enables us to pinpoint issues before they escalate, allowing for timely interventions and corrections. Ultimately, this thorough assessment will enhance the overall efficiency, productivity, and success of the organization, setting a solid foundation for sustained growth and innovation.

1. Align Overall Strategy

In the journey of life and business, one fundamental truth remains: disorder is ever-present and constantly increasing. Recognizing this reality is essential as you face the myriad challenges that come your way. To navigate through these obstacles effectively, you must learn to make good, strategic decisions by following the PATHS METHOD: prioritize, actualize, theorize, harmonize, and summarize.

To begin, your first step is to **prioritize**. This involves carefully assessing the situation at hand and determining which aspects require immediate attention and action. By doing so, you can effectively allocate your resources and energy where they are most needed.

In addition to prioritizing, it is crucial to **actualize** your plans and goals. Mere thoughts and ideas are insufficient; you must take concrete action to turn your visions into reality. This requires determination, perseverance, and a proactive approach to overcome obstacles and achieve success.

While actualizing is essential, it is equally important to **theorize**. This involves analyzing and evaluating the situation from both practical and theoretical perspectives. By deeply understanding the underlying principles and dynamics at play, you can make informed decisions and devise effective strategies.

Harmonizing is another crucial aspect of managing disorder. Life is complex and multifaceted, often requiring various elements to coexist and work together harmoniously. This entails finding the right balance and synergy between different aspects, such as work and personal life, relationships and responsibilities, and short-term and long-term goals.

Lastly, **summarizing** allows you to reflect on your progress and lessons learned. By summarizing your experiences and outcomes, you gain valuable insights that can inform your future actions. This knowledge empowers you to continuously adapt, improve, and navigate through the ever-changing landscape of disorder.

Remember, disorder is an inevitable part of life. However, by prioritizing, actualizing, theorizing, harmonizing, and summarizing, you can effectively manage and overcome the challenges it presents. Stay focused, remain adaptable, and never forget that with the right mindset and strategies, you have the power to transform disorder into growth and success.

2. Align Organizational Structure and Design

In the pursuit of sustainable growth, establishing a robust system for recognizing and engaging your team is essential. This crucial phase focuses on fostering active involvement among team members while aligning with the overarching organizational structure and thoughtfully selected growth strategy.

The first step in this phase involves a thorough examination of role requirements and key performance indicators for each business unit, along with identifying the individuals who fulfill essential roles and functions. By assessing these elements, the organization can pinpoint any talent gaps within the current structure, allowing for prompt remediation.

Additionally, this phase will prioritize the hiring sequence, ensuring that new employees align seamlessly with the company's growth strategy. This meticulous approach to detail will ultimately lead to the development of a highly efficient organizational structure that effectively supports and propels the company toward sustainable growth.

3. Align Organizational Management Activities and Processes

To ensure the effective execution of business strategies, it's essential to implement a structured approach that facilitates both short-term and long-term objectives. This step can be completed on-site and focuses on ensuring the management team is actively working towards the outcomes established in earlier phases, including strategy, key performance indicators (KPIs), and both short-term and mid-term goals.

This may require collecting the right metrics, streamlining data collection and reporting, and providing additional support to the leadership team as they transition into their new roles. It's crucial to manage long-term business development goals separately from these short-term tasks. By establishing distinct processes for short-term and long-term goals, organizations can enhance team decision-making, prioritize effectively during periods of change, and implement necessary adjustments more rapidly.

Adopting this method allows companies to coordinate their immediate and future goals effectively, thereby improving overall performance and flexibility. This also emphasizes the need for accurate data collection and a streamlined evaluation process that supports well-informed decision-making. Furthermore, dedicating time to developing the leadership team's understanding and skills in their new roles provides a strong foundation for success. Separating the management of long-term business development ensures that sustained progress and strategic alignment are maintained without letting short-term tasks overshadow critical long-term objectives.

As a result, the organization's capacity to make collective decisions improves significantly, enabling them to navigate change effectively and capitalize on emerging opportunities. Implementing this comprehensive approach fosters a unified and adaptable business environment, paving the way for significant growth and success.

4. Align Goals, Missions, Targets and Rewards

In step 4, it's essential to acknowledge the strong link to step 3, where collaboration with the leadership team plays a pivotal role in shaping the organization's financial landscape. This step involves working closely with leaders to determine the budget and establish financial targets for the growing business, aligning these elements with the new strategy, structure, and roles.

This collaborative process fosters increased transparency and motivation throughout the organization. By engaging the leadership team in setting financial goals, you create a sense of ownership and commitment to the organization's success. Furthermore, this comprehensive approach enables regular monitoring and assessment of both financial and operational performance, ensuring that the organization remains aligned with its strategic objectives while adapting to any changes in the business environment.

5. Align Culture and People

As we move into this pivotal phase, it's important to recognize how the groundwork laid in previous stages has prepared the senior leadership team for success. This crucial step is about uniting the foundational principles and pillars that underpin the entire organization. By collaborating closely with your leadership team, you will implement the carefully crafted strategy, visionary aspirations, cherished values, goals, mission, roles, and expectations across the company.

The primary objective is to cultivate a shared understanding that clarifies each individual's essential role in shaping the company's destiny, pursuing its vision, and upholding its values. This concerted effort will foster a strong and adaptable organizational culture, reinforcing overall momentum and ensuring success at every stage of the company's journey.

6. Improvements and Adjustments

As we transition into this next phase, the focus will shift towards optimizing the overall performance of the organization as a whole. Interestingly, and perhaps ironically, the dedication and effort invested in achieving proper alignment significantly reduces the need for extensive optimization. With alignment in place, the organization begins to build momentum, steadily progressing down the slope of success propelled by its own inertia. The outcome is a high-performing and resilient organization capable of executing with remarkable power and effectiveness while flawlessly implementing its thoughtfully chosen growth strategy.

---

56. Cooper, Elizabeth. "Cash Crunch: The Silent Killer Of Business." *LinkedIn*, 7 September 2023, https://www.linkedin.com/pulse/cash-crunch-silent-killer-business-elizabeth-cooper/. Accessed 13 October 2024.
57. Porter, Brittany. "Accept Or Decline? How To Evaluate And Respond To Opportunities Strategically." *Forbes*, 31 October 2023, https://www.forbes.com/sites/britneyporter/2023/10/31/accept-or-decline-how-to-evaluate-and-respond-to-opportunities/. Accessed 13 October 2024.
58. Rosenthal, Melissa. "Why does your company exist?" *Medium*, 16 July 2020, https://melissarosenthal.medium.com/why-does-your-company-exist-1a46546df983. Accessed 13 October 2024.
59. Stanford Seed. "Workplace Friction: How to Make the Right Things Easier and the Wrong Things Harder." *Stanford Graduate School of Business*, 13 December 2023, https://www.gsb.stanford.edu/insights/workplace-friction-how-make-right-things-easier-wrong-things-harder. Accessed 13 October 2024.
60. "Top 7 Reasons Why Innovation Fails in Organizations." *Ideapoke*, https://www.ideapoke.com/growthleader/reasons-innovation-fails/. Accessed 13 October 2024.

Conclusion and Next Steps

Congratulations on completing your journey through *Mindset Mechanics*! We've provided additional resources to put you into the driver's seat of success and help you to become a vital leader and happy-high achiever. I hope this book has inspired you to harness the power of your mindset, align your organizational structures, and implement effective strategies that drive success. Remember, the journey doesn't end here; it's just the beginning of your transformation.

If you're ready to take your business to the next level, I invite you to connect with me. Whether you're looking for guidance and more **training materials, online learning, or personalized one-on-one support with private coaching**, I'm here to help you. To enroll in the next **Live Event** or to learn more about my coaching services, contact me at themindsetmechanic.com. Together, we can unlock the full potential of your organization and create a pathway to sustained success.

Thank you for letting me be a part of your journey; on the next page are my contact information and social media handles. Feel free to reach out, and I look forward to connecting with you soon!

Contact and Connect for Coaching Services

Mindset Mechanics, LLC

support@carolettewright.com

337-202-5217

@themindsetmech
@themindsetmech
@themindsetmech
@themindsetmech
@themindsetmech

Linkedin - https://www.linkedin.com/in/themindsetmech/
X (Twitter) - https://x.com/themindsetmech/
Youtube - https://www.youtube.com/@themindsetmech
Instagram - https://www.instagram.com/themindsetmech/
Facebook - https://www.facebook.com/themindsetmech/

*Please send typos and errors to support@carolettewright.com

Made in the USA
Monee, IL
12 November 2026

0220e31c-eb79-4149-b522-b728f8eaa790R01